Microplastics in Focus

Rey Y. Capangpangan • Arnold C. Alguno

Microplastics in Focus

Science, Impact and Action

Springer

Rey Y. Capangpangan
Department of Physical Sciences and Mathematics
Mindanao State University at Naawan
Naawan, Philippines

Arnold C. Alguno
Department of Materials and Resources Engineering and Technology
MSU-Iligan Institute of Technology
Iligan City, Philippines

ISBN 978-3-032-28977-3 ISBN 978-3-032-28978-0 (eBook)
https://doi.org/10.1007/978-3-032-28978-0

This Springer imprint is published by the registered company Springer Nature Switzerland AG
The registered company address is: Gewerbestrasse 11, 6330 Cham, Switzerland

Preface

Plastic has become one of the defining materials of the modern age. Its durability, versatility, and low cost have enabled remarkable advances in healthcare, food preservation, transportation, and technology. Yet these same properties have also created one of the most persistent environmental challenges of our time. As plastics fragment through environmental weathering, they generate microplastics and even smaller particles that now permeate oceans, rivers, soils, the atmosphere, food systems, and living organisms. What was once considered a waste management problem has evolved into a complex scientific, environmental, and societal issue that demands urgent and coordinated action.

Microplastics in Focus: Science, Impact, and Action was conceived to provide a comprehensive, science-grounded, and solution-oriented understanding of this emerging global challenge. This book brings together current knowledge on the formation, distribution, detection, impacts, and mitigation of microplastics, while emphasizing that addressing microplastic pollution requires not only scientific understanding but also governance innovation, technological advancement, and societal engagement.

This book is organized around three interconnected themes. The first part, *The Science of Microplastics*, examines the fundamental processes that define microplastic research, including their origins from material life cycles and degradation processes, their transport and fate across environmental systems, and the analytical methods used to detect and characterize them. Together, these chapters highlight how microplastic research sits at the intersection of materials science, environmental science, and analytical chemistry.

The second part, *The Impact of Microplastics*, explores the biological, ecological, and human dimensions of microplastic exposure. It synthesizes emerging evidence on toxicological mechanisms, ecosystem disruptions, and socioeconomic implications, emphasizing that microplastic pollution is not only an environmental issue but also a food security, public health, and environmental justice concern. By framing microplastics within the broader context of planetary health, this section underscores the interconnected nature of environmental risks in the Anthropocene.

The third part, *Taking Action*, shifts the discussion from understanding the problem to advancing solutions. These chapters demonstrate how effective responses depend on aligning policy instruments, technological innovation, circular economy approaches, and community participation. Importantly, this section highlights that solutions must be systemic, inclusive, and science-informed, reflecting the reality that sustainable mitigation can only be achieved through coordinated efforts among governments, industries, researchers, and communities.

This book also reflects an important transition in the field. Microplastic research is moving beyond documenting contamination toward understanding consequences and designing interventions. Increasingly, the focus is not only on where microplastics are found, but also on what they mean for ecosystems, economies, and societies, and how science can support evidence-based decision-making. By integrating scientific evidence with governance perspectives and practical solutions, this volume aims to contribute to this transition from awareness to action.

The preparation of this book benefited from the collective insights of researchers working across disciplines, including environmental chemistry, marine science, toxicology, materials science, environmental policy, and sustainability science. Such interdisciplinary collaboration reflects the nature of the microplastic challenge itself, which cannot be addressed within disciplinary boundaries alone.

This book is intended for researchers, graduate students, environmental practitioners, policymakers, and development stakeholders seeking a structured and accessible synthesis of microplastic science and its implications. This book will serve not only as a reference for current knowledge but also as a catalyst for future research, innovation, and policy development.

Ultimately, microplastics represent more than an environmental contaminant. They symbolize the unintended consequences of material progress and the urgent need to rethink how society designs, uses, and manages materials. Addressing this challenge requires not only better technologies and stronger policies but also a shared commitment to stewardship, responsibility, and sustainability.

We hope this book contributes, in its own way, to advancing that collective effort.

Naawan, Philippines Rey Y. Capangpangan
Iligan City, Philippines Arnold C. Alguno

Competing Interests

The authors have no competing interests to declare that are relevant to the content of this manuscript.

Contents

List of Figures

About the Authors

Rey Y. Capangpangan is an environmental chemist, researcher, and educator specializing in microplastic pollution, environmental nanotechnology, and catalysis for sustainability applications. As a Professor at Mindanao State University at Naawan (MSUN), Philippines, he leads interdisciplinary research on the detection, environmental fate, risk assessment, and mitigation of emerging contaminants, particularly microplastics in aquatic environments and food systems. His work also advances the development of functional nanomaterials and catalytic technologies for environmental remediation, pollution control, and circular resource utilization. Through his research leadership, international collaborations, and academic mentorship, he promotes science-based, interdisciplinary solutions that connect environmental chemistry, materials science, and engineering to advance sustainable development and protect marine and coastal ecosystems.

Arnold C. Alguno is a Professor in the Department of Materials and Resources Engineering and Technology at the Mindanao State University-Iligan Institute of Technology (MSU-IIT), specializing in semiconductor physics, solar cell technology, and nanotechnology. He obtained his Ph.D. in Physics at Tohoku University, Japan. His research spans a broad range of fields, including sensor technology, quantum dots, and advanced functional materials.

Chapter 1
Introduction: Setting the Stage

1.1 The Age of Plastic: A Boon and a Curse

The unprecedented rise of plastic materials in the past century has revolutionized modern society. Engineered for versatility, durability, and low production cost, plastics now permeate every aspect of daily life, including packaging, transportation, construction, healthcare, agriculture, and advanced electronics. Their chemical stability, light weight, and resistance to degradation enabled large-scale industrial adoption and reshaped global manufacturing systems. Plastics supported food preservation, sterile medical devices, and modern logistics, contributing significantly to improvements in quality of life and economic development [1, 2].

Global plastic production has expanded at an extraordinary rate since the 1950s, reflecting increasing dependence on polymer-based materials. Over 9.2 billion tonnes of plastic have been produced globally, with nearly 80% accumulating in landfills or entering terrestrial and aquatic environments as unmanaged waste [3]. This accumulation reflects linear consumption patterns characterized by production, use, and disposal, with limited recovery and recycling efficiency. Current waste management systems remain insufficient to handle the magnitude and persistence of plastic waste streams [4].

Plastic materials were once regarded as symbols of technological progress. Today, they represent a defining environmental challenge. Environmental exposure transforms large plastic debris into progressively smaller fragments through photodegradation, oxidation, thermal stress, and mechanical abrasion. These processes generate microplastics and nanoplastics that remain persistent in ecosystems due to their resistance to biological degradation [5].

As plastic fragments decrease in size, environmental risks increase. Smaller particles have higher surface area-to-volume ratios, increasing their capacity to adsorb toxic substances and interact with biological systems. This transformation shifts the

R. Y. Capangpangan, A. C. Alguno, *Microplastics in Focus*,
https://doi.org/10.1007/978-3-032-28978-0_1

plastic pollution problem from visible debris to microscopic contaminants that move through air, water, soil, and food systems [6].

Microplastic pollution now represents one of the defining environmental challenges of the twenty-first century. Scientific literature increasingly describes plastics as a planetary boundary issue due to their persistence, global transport, and ecological impacts. Their presence in remote regions such as polar ice, deep-sea sediments, and high-altitude environments confirms their global distribution [5].

The environmental costs associated with plastic production extend beyond waste accumulation. Plastic manufacturing relies heavily on fossil fuel extraction, contributing to greenhouse gas emissions and resource depletion. Projections suggest plastics may account for a significant fraction of global fossil fuel demand by 2050 if current trends continue [5]. These linkages position plastic pollution within broader sustainability and climate discussions.

Despite these challenges, plastics remain indispensable in modern society. Medical devices, water purification systems, renewable energy components, and communication technologies rely heavily on polymer materials. This dual reality creates a complex dilemma. Society depends on plastics while confronting their environmental consequences.

This paradox defines the modern plastic era. Plastics enabled technological advancement while simultaneously creating long-term environmental liabilities. Addressing this contradiction requires scientific innovation, policy reform, circular economy strategies, and behavioral change [1].

Understanding this dual role of plastics provides the foundation for examining microplastics. These particles represent the long-term environmental legacy of plastic dependence. Their emergence signals a transition from visible pollution to invisible contamination, which demands new scientific approaches.

This book begins within this context. It frames microplastics as both a symptom of unsustainable production systems and a scientific frontier requiring interdisciplinary investigation.

1.2 What Are Microplastics? Definitions and Origins

Microplastics are generally defined as synthetic polymer particles measuring less than 5 mm in diameter. This widely accepted definition provides a functional classification based on size rather than composition or origin [5].

Microplastics fall into two major categories. Primary microplastics are intentionally manufactured in microscopic sizes for use in products, such as cosmetics, industrial abrasives, and polymer resins. Secondary microplastics originate from the degradation of larger plastic items through environmental exposure and physical stress [4]. Figure 1.1 illustrates the size classification of plastic debris, the distinction between primary and secondary microplastics, and the major environmental pathways through which these particles enter aquatic, terrestrial, and atmospheric systems.

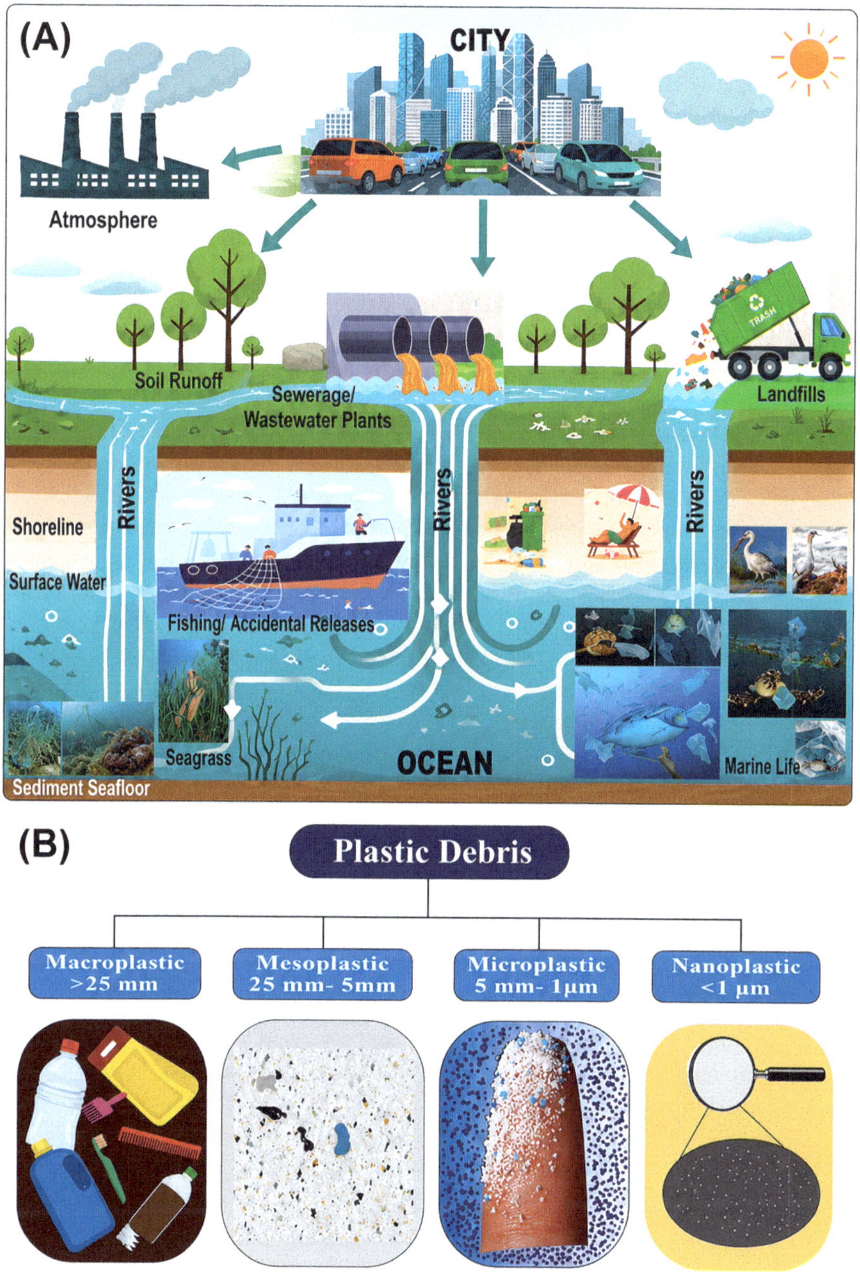

Fig. 1.1 Source, environmental pathways, and size classification of plastic debris and microplastics. (**a**) Conceptual illustration of the major sources and transport pathways through which plastic debris enters aquatic environments, including urban activities, industrial emissions, wastewater effluents, landfill leakage, soil runoff, and atmospheric deposition. The diagram shows the movement of plastics through river systems into coastal and marine ecosystems and their distribution

These particles exhibit significant heterogeneity. Variations occur in polymer type, density, morphology, crystallinity, and chemical additives. Common polymers include polyethylene, polypropylene, polystyrene, polyethylene terephthalate, and polyvinyl chloride. Each polymer exhibits distinct environmental behavior based on its density and degradation characteristics [2].

Environmental fragmentation processes include ultraviolet radiation, temperature fluctuations, hydrolysis, and mechanical abrasion. These forces progressively weaken polymer structures, leading to fragmentation rather than complete mineralization. As a result, plastics rarely break down in the environment. They transform into smaller particles that remain chemically stable [6].

Microplastics also function as environmental carriers of chemical contaminants. Plastic additives, such as plasticizers, flame retardants, and stabilizers, may leach into the surrounding environment. Their hydrophobic surfaces also adsorb persistent organic pollutants and heavy metals [4]. This dual role as both a pollutant and a carrier of contaminants heightens environmental concern. Microplastics represent complex mixtures of polymers and associated chemicals rather than inert particles. Scientific attention toward microplastics accelerated after the early 2000s, when researchers began documenting their widespread occurrence in marine systems. Subsequent studies confirmed their presence in freshwater, soils, atmosphere, and food products [9].

Nanoplastics represent an emerging frontier within this field. These particles, often defined as smaller than 100 nm, exhibit distinct behaviors because they can cross biological membranes and cellular barriers. Their detection remains challenging due to analytical limitations [6].

Understanding the origins of microplastics requires examining the entire life cycles of plastic products. Sources include packaging waste, textile fibers, tire wear particles, fishing gear, construction materials, and wastewater discharges. Urban systems represent major entry points due to high consumption and waste generation [9]. Transport pathways further complicate their distribution. Rivers, wastewater systems, atmospheric transport, and ocean currents redistribute microplastics across regional and global scales. These transport mechanisms connect terrestrial activities with marine contamination [5].

This understanding of definitions and origins establishes the scientific foundation for evaluating impacts. Microplastics must be viewed as dynamic environmental contaminants shaped by production systems, consumption behavior, and environmental processes.

Fig. 1.1 (continued) across environmental compartments, such as surface waters, shorelines, seabed sediments, seagrass habitats, and marine organisms, highlighting the land-sea connectivity of plastic pollution. (**b**) Size-based classification of plastic debris showing the fragmentation continuum from macroplastics (>25 mm) and mesoplastics (25–5 mm) to microplastics (5 mm–1 μm) and nanoplastics (<1 μm), emphasizing how physical, chemical, and biological degradation processes progressively transform larger plastics into smaller particles with greater environmental mobility and biological interaction potential. (Panel (**a**) adapted from [7] and panel (**b**) adapted from [8]. This figure is reproduced under the Creative Commons Attribution 4.0 International License (CC BY 4.0) (https://creativecommons.org/licenses/by/4.0/))

1.3 Why Microplastics Matter: A Global and Regional Concern

Microplastics have transitioned from an emerging research topic into a recognized global environmental concern. Their detection across diverse ecosystems demonstrates their ubiquity and persistence. Studies report the presence of microplastics in deep-sea trenches, Arctic ice, freshwater systems, agricultural soils, drinking water, and human tissues [3]. Their widespread occurrence reflects their mobility and resistance to degradation.

A systems perspective of global occurrence, exposure pathways, and impacts is presented in Fig. 1.2. This figure illustrates the major sources, environmental transport pathways, and exposure routes of microplastics linking urban, freshwater, marine, and food systems. Microplastics originating from urban activities, transportation, industrial emissions, and consumer products enter river systems through surface runoff and wastewater discharge. Wastewater treatment plants are shown as critical control points where microplastics can be partially removed but may still be released into receiving waters. The diagram further highlights the transport of microplastics from rivers to coastal and marine environments, where they accumulate in sediments, aquatic organisms, and aquaculture systems. Potential human exposure pathways are also depicted, including drinking water, seafood

Fig. 1.2 Environmental pathways and human exposure routes of microplastics across the land-water-food nexus. (Adopted from [10]. This figure is reproduced under the Creative Commons Attribution 4.0 International License (CC BY 4.0) (https://creativecommons.org/licenses/by/4.0/))

consumption, sea salt, and atmospheric deposition. Together, the figure highlights the connectivity among anthropogenic sources, environmental transport processes, food-chain contamination, and human exposure risks, underscoring the need for integrated source-to-sea management strategies.

Once released, microplastics interact with biological systems. Aquatic organisms ingest microplastics either directly or indirectly through trophic transfer. Laboratory and field studies demonstrate impacts on feeding behavior, reproduction, growth, and metabolic processes [5]. Microplastics also alter microbial communities and sediment properties. Their presence in soils may influence nutrient cycling and plant growth by altering soil structure and microbial interactions [2]. Human exposure pathways include ingestion, inhalation, and dermal contact. Detection of microplastics in lungs, blood, feces, and the placenta has heightened concerns about potential health impacts. Emerging research links exposure to inflammation, oxidative stress, and immune responses [3]. Economic implications further reinforce the importance of this issue. Plastic pollution imposes costs related to environmental cleanup, fisheries losses, tourism decline, and healthcare burdens. These costs emphasize the need for preventive strategies [1].

Policy responses continue to evolve. International discussions on plastic treaties and regional waste management strategies reflect growing recognition of microplastics as transboundary pollutants. Effective responses require coordinated monitoring, regulation, and technological innovation [1].

Microplastics, therefore, occupy a unique position at the intersection of environmental science, toxicology, economics, and governance. Their complexity requires interdisciplinary collaboration and integrated research approaches [4]. Their significance lies in persistence, mobility, and biological interaction. These characteristics distinguish microplastics from conventional pollutants and justify their classification as priority contaminants.

1.4 Focus on the Global South and Southeast Asia: The Understudied Hotspot

Microplastic research remains geographically uneven. While industrialized nations dominate scientific output, many regions experiencing severe pollution remain underrepresented in research.

Countries in the Global South, particularly Southeast Asia, face disproportionate exposure due to rapid urbanization, coastal population growth, and limited waste management capacity. Leakage of plastic waste into rivers and coastal systems remains a persistent challenge [11]. The ASEAN region represents a critical research frontier. Several countries rank among the largest contributors of marine plastic leakage due to high population densities and inadequate waste infrastructure. Despite this, standardized monitoring frameworks remain limited [12].

Major rivers function as transport corridors linking land-based waste to marine environments. The Mekong River contributes a significant amount of plastic to the South China Sea, influencing coastal pollution patterns across multiple countries [13]. Regional studies confirm widespread contamination in rivers, such as the Chao Phraya, Citarum, and Saigon. Pollution levels often correlate with urban density and challenges in waste management [14]. Coastal ecosystems show high microplastic accumulation. Studies report contamination in sediments, surface waters, and atmospheric fallout [15]. Blue carbon ecosystems, such as mangroves and seagrass beds, also exhibit significant accumulation, raising concerns about ecosystem services [16].

Microplastics also appear in drinking water supplies across Southeast Asia. These findings highlight potential exposure risks for rapidly growing urban populations [17]. Despite growing evidence, regional knowledge gaps remain. Research distribution remains uneven across countries, with limited long-term monitoring programs. Harmonized methodologies remain necessary for comparable datasets [12]. Addressing these gaps requires capacity building, regional collaboration, and investment in analytical infrastructure. Strengthening regional research networks will support policy development and environmental management. Southeast Asia, therefore, represents both a pollution hotspot and a research opportunity. Understanding microplastic dynamics in this region provides insights relevant to other rapidly developing regions.

This book gives particular attention to these geographic disparities. It emphasizes the need for inclusive scientific participation and region-specific solutions.

1.5 Objectives of the Book

This book aims to provide a comprehensive, interdisciplinary, and solution-oriented understanding of microplastic pollution by integrating scientific evidence, societal implications, and actionable responses that support informed decision-making and sustainable environmental management. It establishes a strong scientific foundation by explaining the origin, formation, environmental pathways, detection methods, and environmental behavior of microplastics across terrestrial, aquatic, and atmospheric systems, while clarifying how material design, environmental processes, and waste management limitations collectively drive their proliferation. This book synthesizes current knowledge on the ecological consequences of microplastics, including their impacts on biodiversity, soil systems, food webs, and ecosystem functioning, with particular attention to vulnerable regions, such as the Global South and Southeast Asia. It also examines emerging evidence on human exposure pathways, toxicological mechanisms, and potential long-term health implications, highlighting the need for improved exposure assessment and stronger integration between environmental science and public health research. Beyond the scientific dimension, this book explores the economic and social implications of microplastic pollution, including its influence on production systems, waste governance, social equity, and the disproportionate risks faced by vulnerable communities. It further evaluates governance frameworks, international agreements, and regulatory strategies, such as Extended Producer Responsibility and circular economy approaches, that aim to reduce plastic leakage and prevent microplastic formation. In parallel, this book highlights technological and scientific innovations, including advances in materials science, waste management technologies, circular design strategies, and remediation approaches that support sustainable materials management. It also emphasizes the importance of community engagement through citizen science, environmental education, youth participation, and behavioral change as critical complements to technological and policy interventions. By bridging environmental science, engineering, economics, governance, and public health, this book advances integrated and interdisciplinary approaches that move beyond fragmented responses toward systemic solutions. It also identifies critical knowledge gaps, research priorities, and innovation pathways to guide future scientific inquiry, policy development, and global environmental strategies. Through this integrated perspective, this book advances the discussion of microplastics from problem recognition to coordinated, evidence-based action by connecting science, policy, technology, and society and framing microplastic pollution as a systemic environmental challenge requiring responses across disciplines, sectors, and scales.

References

1. Sin, L. T., & Tueen, B. S. (2024). *Microplastics pollution and worldwide policies on plastic use*. CRC Press. ISBN: 9781032482026. https://doi.org/10.1201/9781003387862
2. Surampalli, R. Y., Zhang, T. C., Al-Hashimi, B. M., Kao, C. M., Ghanrekar, M. M., Bhunia, P., & Das, S. (2025). *Microplastics in the environment: Fate, impacts, removal, and management*. Wiley. ISBN: 9781394251087.
3. Wang, Z., Li, N., Ding, Y., Li, N., Su, M., Zhang, C., Li, Y., Wang, Q., Sha, C., Xia, B., Cheng, J., & Jiang, G. (2025). Microplastics and human health: Exposure pathways, toxicity mechanisms, and future research challenges. *Journal of Environmental Chemical Engineering, 13*, 118807. https://doi.org/10.1016/j.jece.2025.118807
4. Chaudhary, G. R., Chauhan, M., & Sharma, B. (2025). *Microplastics in environment: Source, potential risks, analytical challenges, and remediation strategies*. CRC Press. ISBN: 9781032742045. https://doi.org/10.1201/9781003468141
5. Parmar, S., Sharma, V. K., & Singh, V. (2023). *Microplastics in marine ecosystem: Sources, risks, mitigation technologies, and challenges*. CRC Press. ISBN: 9781032319322. https://doi.org/10.1201/9781003312086
6. Vithanage, M., & Prasad, M. N. V. (2023). *Microplastics in the ecosphere: Air, water, soil, and food*. Wiley & Sons Ltd. ISBN: 9781119879503. https://doi.org/10.1002/9781119879534
7. Bel Hassen, M., Bellaaj Zouari, A., Abdennadher, M., Assaf, J.-C., Nakad, M., Abboud, R., Khammeri, Y., Banni, M., Panzeri, A., Gomes, L., & Hamd, W. (2025). Plastics pollution: Pathways, impacts, and regulatory challenges in marine environments. *Frontiers in Environmental Science, 13*, 1635230. https://doi.org/10.3389/fenvs.2025.1635230
8. Yarahmadi, A., Heidari, S. M., Sepahvand, P., Afkhami, H., & Kheradjoo, H. (2024). Microplastics and environmental effects: Investigating the effects of microplastics on aquatic habitats and their impact on human health. *Frontiers in Public Health, 12*, 1411389. https://doi.org/10.3389/fpubh.2024.1411389

9. Ni, B. J., Xu, Q., & Wei, W. (2023). *Microplastics in urban water management*. John Wiley & Sons, Inc. ISBN: 9781119759348. https://doi.org/10.1002/9781119759379
10. Jiang, B., Kauffman, A. E., Li, L., McFee, W., Cai, B., Weinstein, J., Lead, J. R., Chatterjee, S., Scott, G. I., & Xiao, S. (2020). Health impacts of environmental contamination of micro- and nanoplastics: A review. *Environmental Health and Preventive Medicine, 25*, 29. https://doi.org/10.1186/s12199-020-00870-9
11. Ng, C. H., Mistoh, M. A., Teo, S. H., Galassi, A., Ibrahim, A., Sipaut, C. S., Foo, J., Seay, J., Taufiq-Yap, Y. H., & Janaun, J. (2023). Plastic waste and microplastic issues in Southeast Asia. *Frontiers in Environmental Science, 11*, 1142071. https://doi.org/10.3389/fenvs.2023.1142071
12. Nakano, K., Alfonso, M. B., Phinchan, N., Jandang, S., Abdull Manap, M. R., Chavanich, S., Viyakarn, V., Müller, M., Wong, C., Bacosa, H. P., Celik, M., Cordova, M. R., & Isobe, A. (2025). Aquatic microplastics research in the ASEAN region: Analysis of challenges and priorities. *Marine Pollution Bulletin, 210*, 117342. https://doi.org/10.1016/j.marpolbul.2024.117342
13. Nguyen, D. M., Hole, L. R., Breivik, Ø., Nguyen, T. B., & Pham, N. K. (2023). Marine plastic drift from the Mekong River to Southeast Asia. *Journal of Marine Science and Engineering, 11*(5), 925. https://doi.org/10.3390/jmse11050925
14. Babel, S., Ta, A. T., Nguyen, T. P. L., Sembiring, E., Setiadi, T., & Sharp, A. (2022). Microplastic pollution in selected rivers of Southeast Asia. *APN Science Bulletin, 12*(1), 5–17. https://doi.org/10.30852/sb.2022.1741
15. Rahim, N. N. A., Peng, P. W. Y., Shahrir, N. F., Mahiyuddin, W. R. W., Zain, S. M. S. M., & Ismail, R. (2025). Characteristics, distribution, and sources of atmospheric microplastics in Southeast Asia: A scoping review. *Atmosphere, 16*(5), 515. https://doi.org/10.3390/atmos16050515
16. Adnan, A. A., Iskandar, A., & Arina, N. (2025). Microplastic contamination in Southeast Asia's blue carbon habitats—Systematic review paper with bibliometric approach. *International Journal of Environmental Science and Technology, 22*, 17425–17446. https://doi.org/10.1007/s13762-025-06763-0
17. New, W. X., Kristanti, R. A., Manik, H., Wijayanti, Y., & Adeyemi, D. A. (2023). Occurrence of microplastics in drinking water in Southeast Asia: A short review. *Tropical Environment, Biology, and Technology, 3*(1), 14–24. https://doi.org/10.53623/tebt.v1i1.221

Part I
The Science of Microplastics

Chapter 2
Origin and Formation

Introduction

Plastic pollution did not begin as an environmental problem. It began as a materials innovation success story. Synthetic polymers transformed manufacturing, food systems, medicine, and global trade. Their durability made modern lifestyles possible. That same durability now defines one of the most complex environmental challenges of the century.

Microplastics represent the environmental legacy of this material revolution. They are not accidental pollutants. They are the predictable outcome of production systems designed without end-of-life considerations. Their formation reflects the intersection of polymer chemistry, environmental exposure, and failures in waste management. Understanding the origin and formation of microplastics, therefore, requires a systems perspective. Plastics transition from engineered materials to environmental particles through degradation, fragmentation, transport, and transformation. Each stage influences their environmental behavior and biological interactions.

This chapter establishes the scientific foundation of microplastic formation. It explains how plastics degrade, how sources are classified, how lifecycle transitions shape pollution pathways, and how continued fragmentation produces nanoplastics. These processes explain why microplastics are persistent, mobile, and increasingly difficult to manage. This foundation prepares the reader for subsequent chapters that examine transport, distribution, detection, and impact. Understanding origin is essential because every mitigation strategy depends on understanding how microplastics form.

R. Y. Capangpangan, A. C. Alguno, *Microplastics in Focus*,
https://doi.org/10.1007/978-3-032-28978-0_2

2.1 From Macro to Micro: The Breakdown of Plastics

Plastics are designed to resist degradation. Their chemical stability, hydrophobic structure, and resistance to microbial attack allow them to function under demanding conditions. These same properties explain their persistence after disposal. Once plastics enter the environment, degradation begins immediately. This process rarely leads to complete decomposition. Instead, it produces fragmentation. Large plastic debris gradually transforms into smaller particles through environmental exposure.

Macroplastics are plastic debris larger than 5 mm. These include packaging films, fishing nets, containers, bottles, and synthetic fabrics. Environmental forces progressively transform these materials into microplastics [1]. Solar radiation represents one of the most important degradation drivers. Ultraviolet exposure initiates photooxidation reactions that weaken polymer chains. Polyethylene and polypropylene are particularly vulnerable due to their widespread use and exposure patterns [2]. Photooxidation initiates chain-scission reactions that reduce tensile strength. As polymers lose structural cohesion, surface cracks develop. These cracks become fragmentation points under mechanical stress.

Mechanical degradation accelerates this process. Wave action, sediment abrasion, and wind-driven friction continuously stress weakened materials. Coastal and river environments, therefore, function as active fragmentation zones [3]. Temperature variation also contributes to degradation. Repeated heating and cooling cycles induce thermal stress, promoting polymer oxidation and embrittlement [4]. Biological degradation plays a minor role in conventional plastics. Some microorganisms metabolize additives or partially oxidized polymers. Complete biodegradation remains rare. Most plastics fragment rather than mineralize [5].

Figure 2.1 illustrates the major primary and secondary sources of microplastics and the key environmental weathering processes that contribute to their formation and transformation. Primary microplastics originate from personal care products and industrial applications, while secondary microplastics arise from the fragmentation of larger plastic debris. The diagram further highlights the roles of environmental drivers, including ultraviolet (UV) radiation, biological degradation, wave action, and temperature fluctuations, in accelerating the aging, fragmentation, and physicochemical transformation of plastics. Together, these processes influence particle size reduction, surface modification, and environmental behavior, emphasizing the importance of weathering mechanisms in controlling the generation, transformation, and persistence of microplastics in natural environments.

Primary microplastics originate from intentionally manufactured small particles used in personal care and industrial products, whereas secondary microplastics are generated through the progressive fragmentation of larger plastic debris. As shown in the figure, environmental drivers such as ultraviolet (UV) radiation, biological degradation, wave action, and temperature fluctuations contribute to the aging and breakdown of plastic materials. These processes not only reduce particle size but also modify surface structure and chemical composition, transforming plastics into more environmentally reactive particles with altered environmental behavior.

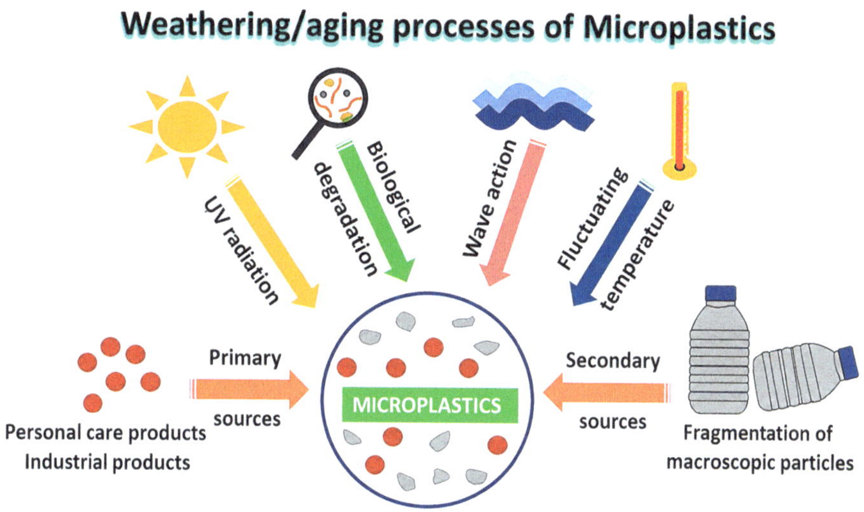

Fig. 2.1 Primary and secondary sources and environmental weathering processes governing microplastic formation. (Adapted from [6]. This figure reproduced under the Creative Commons Attribution 4.0 International License (CC BY 4.0) (https://creativecommons.org/licenses/by/4.0/))

Weathering processes increase surface roughness and introduce oxygen-containing functional groups, thereby enhancing microplastics' ability to adsorb metals and hydrophobic organic contaminants. At the same time, biological interactions such as microbial colonization further modify particle behavior through biofilm formation, which can influence particle density, environmental transport, and ecological interactions. These transformations demonstrate that degradation represents a process of environmental transformation rather than disappearance, with plastics persisting as chemically and biologically interactive particles. Understanding these weathering pathways is therefore essential for explaining the persistence, mobility, and evolving risks of microplastics in the environment.

Secondary microplastics generated through these processes dominate environmental samples. These particles frequently retain additives and sorbed contaminants, increasing their environmental reactivity [7]. Fragmentation also alters surface properties by increasing surface roughness and the presence of oxygen-containing functional groups, which enhance adsorption capacity for metals and hydrophobic contaminants [8]. Biofilm formation further modifies particle behavior as microorganisms colonize plastic surfaces, altering density, transport pathways, and ecological interactions [9].

Degradation, therefore, represents transformation rather than disappearance. Plastics evolve into chemically active environmental particles rather than being completely mineralized. Understanding these processes explains why microplastics persist and why their risks evolve over time. This macro-to-micro transition represents a critical early stage in the development of microplastic pollution because it determines particle size distribution, transport behavior, and exposure potential.

2.2 Categories of Microplastics: Primary Versus Secondary Sources

Microplastics originate from multiple pathways, and their classification into primary and secondary microplastics provides a useful framework for understanding pollution sources and identifying appropriate intervention strategies. Figure 2.2 presents a conceptual overview of this classification, showing primary microplastics as intentionally manufactured particles smaller than 5 mm used in commercial and industrial applications, including microbeads, industrial resin pellets, and synthetic fibers. In contrast, secondary microplastics are formed through the environmental degradation of larger plastic debris, such as plastic bottles, bags, fishing gear, and tire wear particles, driven by processes, including ultraviolet radiation, mechanical weathering, oxidation, and biological activity. Both pathways ultimately generate microplastic particles that are transported across environmental compartments, including water, soil, air, and biological systems.

Primary microplastics are manufactured intentionally in microscopic dimensions. These include polymer pellets used as industrial feedstock, microbeads used in personal care products, blasting abrasives, and engineered polymer powders. Their small size enables specialized industrial applications but also increases environmental mobility [10]. Polymer pellets serve as raw materials for plastic manufacturing. Accidental release during transport and processing contributes to environmental contamination. These particles are frequently found in coastal sediments near industrial areas.

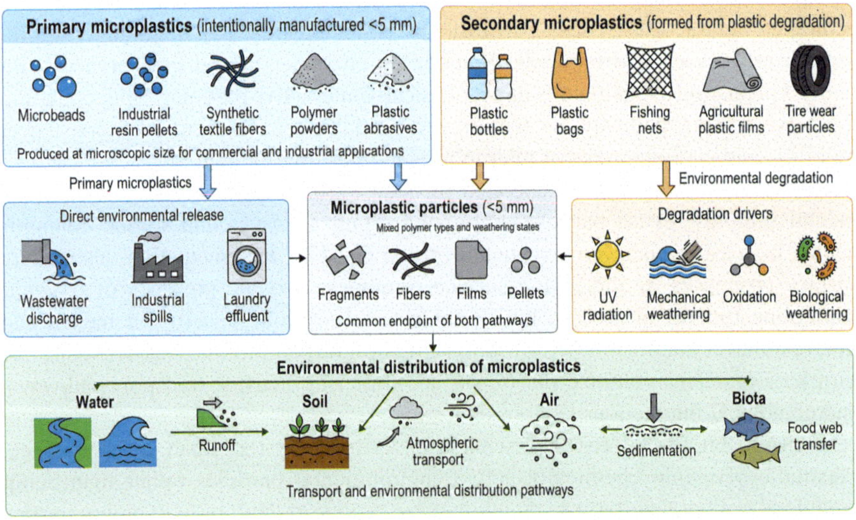

Fig. 2.2 Primary and secondary microplastics—sources, formation pathways, and environmental distribution. (Conceptual framework developed by the authors and illustrated using digital design tools)

Microbeads historically represented a major source due to their use in exfoliating cosmetics. Regulatory actions in several countries have restricted these applications. However, other industrial uses remain widespread and poorly regulated [11]. Synthetic textile fibers represent another major primary source. Washing processes release large quantities of polyester and nylon fibers into wastewater systems. Wastewater treatment plants remove a significant fraction, yet substantial quantities still enter aquatic environments [12].

Secondary microplastics originate from the environmental degradation of larger plastics, as illustrated in Fig. 2.2, where weathering processes progressively transform microplastic debris into microplastic particles. These particles dominate environmental contamination due to extensive plastic consumption and disposal. Environmental conditions strongly influence secondary formation. Ultraviolet exposure, temperature variation, oxygen availability, and salinity all affect degradation rates. Warmer climates and high-radiation environments often exhibit faster rates of fragmentation [13].

Common secondary sources include plastic packaging, agricultural films, fishing equipment, tire wear particles, and construction debris. Tire wear particles represent an especially significant urban source due to continuous abrasion during transportation. Source identification remains technically challenging. Environmental particles often undergo weathering, obscuring their original morphology. Spectroscopic techniques, such as Fourier transform infrared and Raman spectroscopy, support source identification through polymer fingerprinting [14].

Distinguishing sources remains essential for mitigation planning. Primary sources require upstream regulatory interventions. Secondary sources require improvements in waste management, product design, and material recovery systems [15]. Effective management, therefore, depends on understanding source diversity. Pollution prevention strategies must address both intentional production and environmental generation.

2.3 The Life Cycle of Plastics: From Production to Pollution

The life cycle of plastics begins with the extraction of fossil resources and extends through manufacturing, consumption, disposal, and environmental redistribution, reflecting the close relationship between material innovation and environmental consequence. Figure 2.3 presents a simplified conceptual lifecycle framework illustrating how plastics transition from production and use to waste generation, environmental exposure, and progressive fragmentation into microplastics (<5 mm). Environmental weathering processes, including ultraviolet radiation, mechanical stress, oxidation, and biological degradation, drive the breakdown of macroplastics into smaller particles that are subsequently transported through water, soil, air, and biota. Together, these pathways highlight the systemic connections between material lifecycle stages and environmental pollution while also emphasizing opportunities for circular economy interventions to reduce plastic leakage.

Global plastic production exceeded 400 million metric tons annually by 2022. Packaging represents the largest sector, followed by construction, textiles,

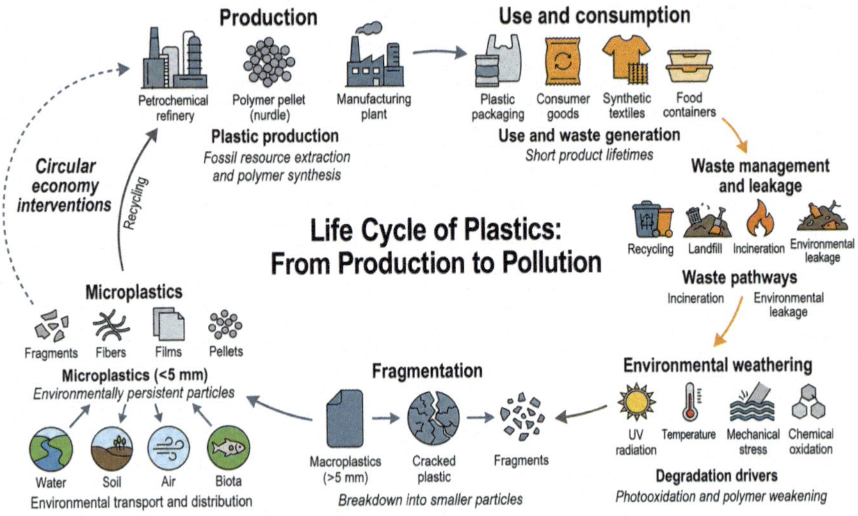

Fig. 2.3 The life cycle of plastics—from production to pollution. (Conceptual framework developed by the authors and illustrated using digital design tools)

automotive components, and consumer goods [16]. Many plastic products serve short functional lifetimes. Packaging materials are often used for less than 1 year. Environmental persistence, however, extends for decades or centuries [13].

Waste management determines environmental fate. Plastics may enter recycling systems, landfills, incineration facilities, or uncontrolled environments. Regions with limited waste infrastructure often experience higher leakage rates [17]. Environmental exposure initiates degradation and fragmentation. Plastics entering rivers, soils, and coastal zones can serve as sources of microplastic formation. These processes create a continuous transition from product use to environmental contamination.

Material composition influences degradation patterns. Thin polyethylene films fragment faster than rigid polyvinyl chloride structures. Additive composition also influences degradation pathways and toxicity profiles [18]. Chemical additives represent an additional concern. Plasticizers, stabilizers, pigments, and flame retardants may leach during degradation. These chemicals contribute to environmental toxicity independent of the plastic particles themselves [19].

Lifecycle assessment studies highlight opportunities for intervention. Circular economy approaches emphasize reduction, reuse, redesign, and recycling. Chemical recycling technologies aim to recover polymer value while reducing environmental leakage [20]. Despite these advances, global management systems remain fragmented. Plastic leakage continues across supply chains due to economic disparities, infrastructure gaps, and regulatory variation.

Understanding lifecycle transitions, as summarized in Fig. 2.3, supports the development of systemic solutions. Microplastics represent the final stage of inefficient material systems. Addressing their formation requires intervention across the entire life cycle rather than relying solely on downstream remediation.

2.4 Link to Nanoplastics: A Continuum of Concern

Microplastics exist along a size continuum that extends into the nanoscale, with particles smaller than 1 μm increasingly recognized for their unique properties and risks. Nanoplastics result from the further fragmentation of microplastics and can also be manufactured directly for specialized applications. Their size allows them to cross biological membranes, interact with cells, and even penetrate the blood-brain barrier in model organisms [21].

Environmental degradation processes, especially mechanical abrasion and photooxidation, lead to progressive size reduction of plastic particles, creating a dynamic spectrum from macro to nano. However, detecting and characterizing nanoplastics remains analytically challenging. Most conventional techniques fail to resolve particles below 1 μm, complicating risk assessment and exposure modeling [22].

The environmental behavior of nanoplastics diverges significantly from that of larger particles. They exhibit colloidal stability, enhanced mobility in soils and water, and higher surface area-to-volume ratios, making them more reactive to contaminants [23]. These properties elevate their potential to bioaccumulate and induce sublethal effects in aquatic and terrestrial organisms [24].

While microplastics have attracted regulatory attention, nanoplastics remain largely unregulated due to methodological limitations. Nonetheless, the transition from micro to nanoplastics is neither discrete nor insignificant. It underscores a continuum of concern where particle size determines not only physical behavior but also biological reactivity and policy urgency. Future research must prioritize detection technologies, exposure modeling, and toxicological evaluation. Addressing nanoplastics requires integration of environmental chemistry, toxicology, and materials science.

2.5 Chapter Synthesis

This chapter establishes a central scientific message. Microplastics do not originate as isolated contaminants. They emerge from material design, consumption systems, and environmental transformation processes.

Four key insights emerge:

First, plastics degrade through fragmentation rather than disappearance.Second, microplastic sources reflect both intentional production and environmental breakdown. Third, lifecycle inefficiencies drive environmental leakage. Fourth, continued fragmentation produces nanoplastics with emerging risks.

These processes explain why microplastics are persistent and globally distributed. They also explain why prevention must focus on material systems rather than downstream cleanup alone.

The next chapter extends this foundation by examining how microplastics move beyond their points of origin and become embedded within environmental systems. It explores the pathways that distribute these particles across terrestrial, freshwater, marine, and atmospheric environments, the processes that govern their persistence and accumulation, and the emerging evidence from global and Southeast Asian pollution hotspots. Understanding the origin explains how microplastics form. Understanding environmental pathways and fate explains how they persist, circulate, and ultimately shape ecological exposure.

Together, these perspectives frame microplastics as a systems-level environmental challenge.

References

1. Cole, M., Lindeque, P., Halsband, C., & Galloway, T. S. (2011). Microplastics as contaminants in the marine environment: A review. *Marine Pollution Bulletin, 62*(12), 2588–2597. https://doi.org/10.1016/j.marpolbul.2011.09.025
2. Gewert, B., Plassmann, M. M., & MacLeod, M. (2015). Pathways for degradation of plastic polymers floating in the marine environment. *Environmental Science: Processes & Impacts, 17*(9), 1513–1521. https://doi.org/10.1039/C5EM00207A
3. Browne, M. A., Galloway, T. S., & Thompson, R. C. (2010). Spatial patterns of plastic debris along estuarine shorelines. *Environmental Science & Technology, 44*(9), 3404–3409. https://doi.org/10.1021/es903784e
4. Mucha, M., & Mucha, J. (2022). Heterogeneous polymers degradation—Microplastic. *Polimery, 67*, 293–297. https://doi.org/10.14314/polimery.2022.7.1
5. Shah, A. A., Hasan, F., Hameed, A., & Ahmed, S. (2008). Biological degradation of plastics: A comprehensive review. *Biotechnology Advances, 26*(3), 246–265. https://doi.org/10.1016/j.biotechadv.2007.12.005
6. Pandey, B., Pathak, J., Singh, P., Kumar, R., Kumar, A., Kaushik, S., & Thakur, T. K. (2023). Microplastics in the ecosystem: An overview on detection, removal, toxicity assessment, and control release. *Water, 15*(1), 51. https://doi.org/10.3390/w15010051
7. Andrady, A. L. (2011). Microplastics in the marine environment. *Marine Pollution Bulletin, 62*(8), 1596–1605. https://doi.org/10.1016/j.marpolbul.2011.05.030
8. Verla, A. W., Enyoh, C. E., Verla, E. N., & Nwarnorth, K. O. (2019). Microplastic-toxic chemical interaction: A review study on quantified levels, mechanism and implication. *SN Applied Sciences, 1*, 1400. https://doi.org/10.1007/s42452-019-1352-0
9. Corcoran, P. L. (2022). *Degradation of microplastics in the environment* (pp. 531–542). Springer eBooks. https://doi.org/10.1007/978-3-030-39041-9_10
10. Frias, J. P. G. L., & Nash, R. (2019). Microplastics: Finding a consensus on the definition. *Marine Pollution Bulletin, 138*, 145–147. https://doi.org/10.1016/j.marpolbul.2018.11.022
11. Han, J. H., & Kim, H. S. (2025). Microplastics in cosmetics: Emerging risks for skin health and the environment. *Cosmetics, 12*(4), 171. https://doi.org/10.3390/cosmetics12040171
12. Raju, S., Carbery, M., Kuttykattil, A., Senathirajah, K., Subashchandrabose, S. R., Evans, G., & Thavamani, P. (2018). Transport and fate of microplastics in wastewater treatment plants: Implications to environmental health. *Reviews in Environmental Science and Bio/Technology, 17*(4), 637–653. https://doi.org/10.1007/S11157-018-9480-3

13. Geyer, R., Jambeck, J. R., & Law, K. L. (2017). Production, use, and fate of all plastics ever made. *Science Advances, 3*(7), e1700782. https://doi.org/10.1126/sciadv.1700782
14. Hernandez, L. M., Yousefi, N., & Tufenkji, N. (2017). Are there nanoplastics in your personal care products? *Environmental Science & Technology Letters, 4*(7), 280–285. https://doi.org/10.1021/acs.estlett.7b00187
15. Thompson, R. C., Swan, S. H., Moore, C. J., & vom Saal, F. S. (2009). Our plastic age. *Philosophical Transactions of the Royal Society B, 364*(1526), 1973–1976. https://doi.org/10.1098/rstb.2009.0054
16. Pathak, P., Sharma, S., & Ramkrishna, S. (2023). Circular transformation in plastic management lessen the carbon footprint of the plastic industry. *Materials Today Sustainability, 22*, 100365. https://doi.org/10.1016/j.mtsust.2023.100365
17. Mustapha, A. (2025). The disproportionate impacts of plastics on low- and middle-income countries. *Journal of Moral Theology, 7(CTEWC Book Series 7).* https://doi.org/10.55476/001c.141262
18. Hale, R. C., King, A., & Ramirez, J. M. (2024). *Plastic debris: An overview of composition, sources, environmental occurrence, transport, and fate* (pp. 1–31). Elsevier BV. https://doi.org/10.1016/b978-0-443-15332-7.00004-1
19. Teuten, E. L., Saquing, J. M., Knappe, D. R. U., Barlaz, M. A., Jonsson, S., Björn, A., Rowland, S. J., Thompson, R. C., Galloway, T. S., Yamashita, R., Ochi, D., Watanuki, Y., Moore, C. J., Viet, P. H., Tana, T. S., Prudente, M. S., Boonyatumanond, R., Zakaria, M. P., Akkhavong, K., … Takada, H. (2009). Transport and release of chemicals from plastics to the environment and to wildlife. *Philosophical Transactions of the Royal Society B: Biological Sciences, 364*(1526), 2027–2045. https://doi.org/10.1098/RSTB.2008.0284
20. Ledakowicz, S., & Zieminska-Stolarska, A. (2023). The role of life cycle assessment in the implementation of circular economy in sustainable future. *Chemical and Process Engineering: New Frontiers, 44*(4), e37. https://doi.org/10.24425/cpe.2023.147396
21. Shukla, S., Khanna, S., & Khanna, K. (2024). Unveiling the toxicity of micro-nanoplastics: A systematic exploration of understanding environmental and health implications. *Toxicology Reports, 14*, 101844. https://doi.org/10.1016/j.toxrep.2024.101844
22. Koelmans, A. A., Besseling, E., & Shim, W. J. (2015). Nanoplastics in the aquatic environment. Critical review. In M. Bergmann, L. Gutow, & M. Klages (Eds.), *Marine anthropogenic litter*. Springer. https://doi.org/10.1007/978-3-319-16510-3_12
23. Kumar, R., Chawla, J., & Syal, J. (2024). Detection, monitoring, and absorption of micro/nano-plastics in soil system. In *Global impacts of micro- and nano-plastic pollution* (Advances in environmental engineering and green technologies) (pp. 77–98). IGI Global. https://doi.org/10.4018/979-8-3693-3447-8.ch004
24. Prata, J. C., da Costa, J. P., Lopes, I., Duarte, A. C., & Rocha-Santos, T. (2020). Environmental exposure to nanoplastics: An overview on possible human health effects. *Science of the Total Environment, 702*, 134455. https://doi.org/10.1016/j.scitotenv.2019.134455

Chapter 3
Microplastics in the Environment

Introduction

The science of microplastics in the environment examines how these synthetic particles enter natural systems, move across environmental compartments, undergo transformation, and persist across spatial and temporal scales. Research shows that most microplastics originate from land-based activities, including urban runoff, tire wear, wastewater discharge, sewage sludge application, and intensive agricultural plastic use. Rivers, drainage networks, coastal circulation, atmospheric transport, and biological vectors then redistribute these particles across freshwater systems, marine environments, soils, and the atmosphere. Their occurrence reflects structural patterns in production, consumption, and waste governance rather than isolated contamination events.

After release, microplastics interact continuously with sediments, organic matter, microbial communities, and co-occurring contaminants. These interactions alter particle density, surface chemistry, transport behavior, and environmental residence time. Such processes determine whether particles remain suspended, settle to the sediment, re-enter the water column, or become incorporated into food webs. Environmental conditions, including ultraviolet radiation, mechanical abrasion, salinity gradients, and temperature fluctuations, further modify particle structure and influence fragmentation and aging.

A systems perspective, therefore, becomes essential. Microplastics move through interconnected environmental pathways linking land, water, and air. Their presence signals how modern material flows intersect with ecological processes. Understanding these pathways establishes the scientific basis for identifying exposure routes, evaluating long-term ecological risks, and guiding mitigation strategies grounded in empirical evidence. This chapter develops that foundation by examining environmental transport processes, environmental fate, and regional pollution patterns, with particular attention to emerging evidence from Southeast Asia.

R. Y. Capangpangan, A. C. Alguno, *Microplastics in Focus*,
https://doi.org/10.1007/978-3-032-28978-0_3

3.1 Pathways: How Microplastics Spread in Land, Water, and Air

Microplastics have permeated every environmental domain, reaching terrestrial soils, aquatic systems, and atmospheric layers through a complex web of interactions between human activity and natural processes. These pathways are neither random nor isolated. They reflect systemic failures in plastic management and reveal intricate connections between production, consumption, and environmental release. Understanding how microplastics travel is crucial for identifying exposure routes, predicting long-term ecological impacts, and designing mitigation interventions.

Land-based sources dominate the global microplastic budget. Urban runoff, improper waste disposal, tire abrasion, sewage sludge application, and agricultural plastic use contribute significantly to the initial environmental release of microplastics into soils and surface waters [1]. Wastewater treatment plants, although partially effective at filtering larger particles, often discharge effluents laden with smaller microplastics into rivers and estuarine zones [2]. These particles can accumulate in agricultural lands when biosolids are applied as fertilizers, thus closing the loop between consumption and terrestrial contamination [3].

Once released, microplastics enter fluvial systems where hydrological dynamics, sediment interactions, and biotic uptake determine their residence time and mobility. Studies have shown that rivers such as the Thames, Danube, and Mekong act as key conduits, funneling massive loads of microplastics into marine environments [4–6]. The surface hydrology, turbulence, and geomorphology of river basins control the deposition and resuspension of microplastics, with larger particles often settling in sediments and finer fractions continuing downstream transport.

In marine systems, microplastics exhibit horizontal and vertical movement influenced by particle density, shape, and biofouling. Buoyant polymers such as polyethylene and polypropylene remain at the surface, facilitating long-range dispersal via ocean currents. Denser polymers, or those colonized by biofilms, sink into the water column and into seabed sediments [7]. These dynamics create stratified distributions of microplastics across coastal, pelagic, and benthic zones. Coastal embayments, estuaries, and coral reefs serve as hotspots of accumulation due to localized retention and reduced hydrodynamic flushing [8].

Atmospheric pathways further amplify the spatial reach of microplastics. Road traffic, synthetic textiles, industrial emissions, and construction activities release airborne fibers and fragments that disperse across urban and remote landscapes alike. Recent studies confirm atmospheric deposition as a non-negligible input route into terrestrial and aquatic environments [9, 10]. Microplastics travel via wind currents, rainout, and dry deposition, potentially entering the hydrological cycle and reintroducing contaminants into surface waters. These transport routes interact dynamically. Atmospheric deposition can return plastics to soils and waters; fluvial systems can re-entrain particles during floods or erosion events; coastal winds can remobilize plastics from beach sediments. Seasonal variability, land use, and meteorological conditions modulate these exchanges. Without a systems-level understanding of these cross-media flows, risk assessments will remain fragmented, and management efforts will be ineffective [11].

Each environmental compartment functions both as a reservoir and as a redistribution node within the planetary microplastic cycle. Addressing microplastic pollution, therefore, requires an integrated perspective that recognizes land, water, and air not as isolated domains but as interconnected interfaces where particles circulate, accumulate, and interact with ecological and human systems. Figure 3.1 illustrates the interconnected pathways through which microplastics are transported from terrestrial sources to coastal and marine environments.

Atmospheric deposition and wind dispersal facilitate the long-range movement of airborne microplastics, allowing particles to settle in both inland and coastal regions. River systems serve as major conduits transporting microplastics from upstream catchments toward densely populated coastal zones. In urban coastal agglomerations, additional inputs arise from anthropogenic activities, including wastewater treatment plant (WWTP) discharges, industrial emissions, and leakage from coastal landfills. Ocean currents subsequently redistribute these particles within marine ecosystems, where they can accumulate in the water column and interact with marine organisms. Together, the figure highlights the dynamic connectivity between atmospheric transport, riverine fluxes, urban emissions, and ocean circulation in shaping the global distribution of microplastics, emphasizing the need for integrated land-sea-atmosphere management strategies.

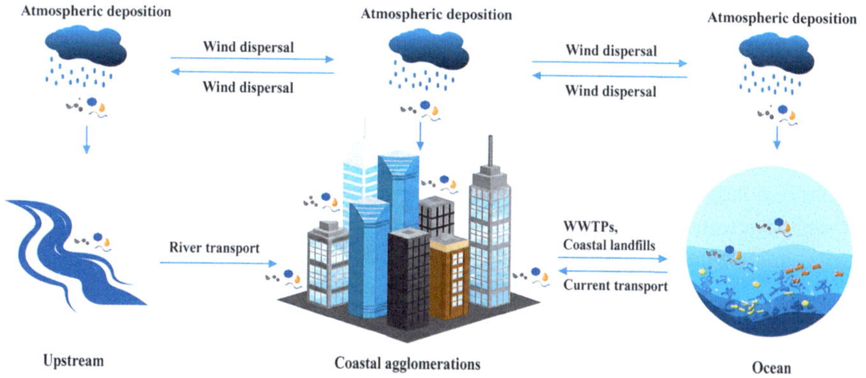

Fig. 3.1 Environmental transport pathways of microplastics across the atmospheric, freshwater, urban, and marine systems. (Adapted from [12]. This figure reproduced under the Creative Commons Attribution 4.0 International License (CC BY 4.0) (https://creativecommons.org/licenses/by/4.0/))

3.1.1 *Pathways Through Land*

Terrestrial ecosystems are among the least studied yet serve as both reservoirs and sources of microplastics [13]. Agriculture contributes significantly through the application of plastic mulches, controlled-release fertilizers encapsulated in polymers, and wastewater irrigation with synthetic fibers [14]. These particles infiltrate the soil matrix, altering soil porosity, microbial activity, and nutrient cycling. Landfills and dumpsites, especially in developing regions, leak fragmented macroplastics into nearby soils via runoff, scavenging animals, or wind dispersal [15]. Biosolids from wastewater treatment plants—used as fertilizers in many countries—introduce millions of microplastic particles per hectare [16].

Road dust, tire wear particles, and atmospheric deposition further enhance the presence of microplastics in soils near urban areas and transport corridors. These inputs are often overlooked in regulatory frameworks, despite their potential to persist and bioaccumulate in terrestrial food webs [17].

3.1.1.1 Terrestrial Entry Points of Microplastics

- Plastic mulching and agrochemicals
- Biosolid application
- Runoff from urban centers and roads
- Improper landfill management

3.1.2 *Pathways Through Water*

Aquatic environments represent the most visible sink for microplastics. Rivers serve as major conduits, channeling plastic debris from land to sea. Plastic particles from urban wastewater, stormwater overflows, and industrial effluents flow into streams, lakes, and ultimately marine systems [6]. In freshwater systems, microplastics accumulate in sediments, especially in slow-moving or enclosed water bodies [18]. Wastewater treatment plants (WWTPs), while efficient at removing >90% of microplastics, still release billions of fibers and fragments daily via effluent and sludge pathways [19]. Fishing gear loss, aquaculture net degradation, and maritime shipping operations also contribute to microplastic loadings in both coastal and offshore waters.

Microplastics can float, suspend, or sink depending on their polymer type, biofilm colonization, and environmental turbulence. These behaviors govern their residence time and spatial distribution, from surface waters to abyssal sediments [20].

3.1.2.1 Freshwater and Marine Pathways

- Surface runoff and urban wastewater
- Industrial discharges
- Fragmentation of fishing gear
- Ocean gyres and long-range transport

3.1.3 Pathways Through Air

Atmospheric transport of microplastics, once underestimated, has gained recognition as a significant pathway of dissemination. Fibers from textiles, tire wear, and construction dust become airborne through mechanical disturbance and wind erosion [9]. Urban atmospheres contain concentrations ranging from hundreds to thousands of particles per cubic meter. Long-range atmospheric transport carries microplastics across continents and into remote ecosystems, such as the Arctic, the Alps, and high-altitude lakes [21]. Indoor air also poses health risks due to high concentrations of synthetic fibers from furniture, carpets, and ventilation systems. Recent studies suggest that precipitation acts as a major deposition mechanism, washing airborne particles into terrestrial and aquatic surfaces [10]. Despite growing evidence, regulatory measures for airborne microplastics remain absent.

3.1.3.1 Atmospheric Microplastic Sources

- Synthetic textiles and household dust
- Construction and industrial emissions
- Resuspension from roads and soils

3.2 Environmental Fate: Persistence and Accumulation in Ecosystems

3.2.1 Persistent Contaminants in a Fragmenting World

Microplastics represent a persistent pollutant class, deeply integrated into modern ecosystems. These synthetic particles resist biodegradation due to their durable polymeric structures, engineered to withstand physical, chemical, and biological weathering. Their long environmental residence time—often extending over centuries—ensures continued accumulation across ecological compartments, from surface waters to deep sediments and terrestrial soils [22, 23]. Unlike many pollutants that degrade into inert forms, microplastics fragment into progressively smaller particles while retaining the original polymer identity, enhancing their environmental mobility and biological interaction potential [24].

3.2.2 *Aquatic Sinks and Circularity Persistence*

Marine and freshwater environments serve as major environmental sinks for microplastics. In surface waters, buoyant polymers such as polyethylene (PE) and polypropylene (PP) accumulate in oceanic gyres through the combined influence of wind-driven currents and large-scale convergence zones. The North Pacific Subtropical Gyre, widely known as the Great Pacific Garbage Patch, exemplifies this process by concentrating microplastic debris at unprecedented densities [20, 25]. Figure 3.2 presents an integrated overview of the major anthropogenic sources, environmental transport mechanisms, and ultimate sinks of microplastics.

Key sources include industrial activities, domestic plastic use, wastewater treatment plants, coastal landfills, personal protective equipment, agricultural plastic mulches, tire wear particles, fisheries, and tourism-related waste. These materials enter the environment through multiple transport pathways, such as atmospheric wet deposition, dust transport, surface runoff, wind dispersal, and riverine transport, facilitating the movement of microplastics across terrestrial, freshwater, and coastal systems. The figure further illustrates the accumulation of microplastics in environmental sinks, including soils, sediments, and terrestrial plants, as well as their uptake by animals and potential transfer to humans through environmental exposure

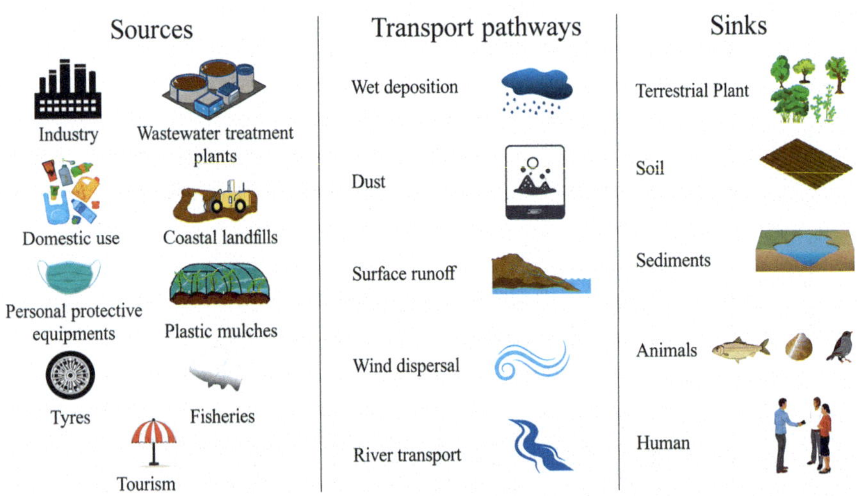

Fig. 3.2 Sources, environmental transport pathways, and sinks of microplastics across terrestrial and aquatic systems. (Adapted from [12]. This figure reproduced under the Creative Commons Attribution 4.0 International License (CC BY 4.0) (https://creativecommons.org/licenses/by/4.0/))

and food web interactions. Together, the diagram highlights the interconnected source-pathway-sink continuum that governs the environmental fate, distribution, and biological exposure risks of microplastics.

Heavier polymers, such as polyethylene terephthalate (PET) and polyvinyl chloride (PVC), and composite particles settle in sediments, particularly in estuarine and coastal zones where hydrodynamic energy decreases. These sediment-bound particles can act as long-term pollutant reservoirs, contributing to chronic contamination and serving as secondary sources through resuspension events during storms or anthropogenic disturbances [26]. In freshwater systems, microplastics often follow a similar fate, accumulating in riverbeds or lakes, where biological activity and particle aggregation influence vertical transport [18].

Microplastics exhibit remarkable transboundary persistence. Hydrological connectivity, such as river outflows, estuarine mixing, and tidal exchange, facilitates redistribution across coastal interfaces, often leading to deposition in biologically rich habitats, such as coral reefs, seagrass beds, and mangroves [27]. These zones become hotspots of contamination, disrupting trophic interactions and degrading ecosystem services.

3.2.3 *Bioaccumulation and Trophic Transfer*

Microplastics interact with aquatic food webs at multiple levels. Phytoplankton, zooplankton, and filter-feeding invertebrates ingest particles directly due to their size similarity with natural prey or by mistake during feeding [4, 8]. Once ingested, particles can impair feeding, growth, and reproduction, while smaller sizes increase potential for translocation across epithelial barriers.

Higher-trophic-level organisms, including commercially important fish and shellfish, accumulate microplastics through dietary exposure and gill uptake, reflecting their position within complex aquatic food webs. Laboratory and field studies have reported gastrointestinal blockage, tissue inflammation, oxidative stress, and biochemical disturbances associated with microplastic ingestion, although sublethal and chronic effects remain insufficiently explored for many species [28]. Beyond direct exposure, the trophic transfer of microplastics has been documented in both marine and freshwater ecosystems, beginning with their entry into planktonic communities and subsequent transfer to small fish, larger predators, and ultimately to humans through seafood consumption [29]. This process highlights the role of microplastics not only as physical contaminants but also as vectors of hydrophobic organic pollutants and plastic additives that can adsorb onto particle surfaces and move across trophic levels.

In the context of bioaccumulation and trophic transfer, smaller organisms such as zooplankton may ingest microplastics due to their size similarity with natural food particles. These contaminated organisms are then consumed by higher trophic species, enabling the progressive movement of microplastics and associated chemicals through the food web. While clear evidence of classical biomagnification remains under scientific debate, repeated ingestion across trophic interactions suggests the potential for cumulative exposure, particularly in long-lived predatory species.

Figure 3.3 illustrates the environmental distribution of microplastics across marine compartments and their transfer through aquatic food webs. Microplastics are found in surface waters, the water column, and benthic sediments, where they

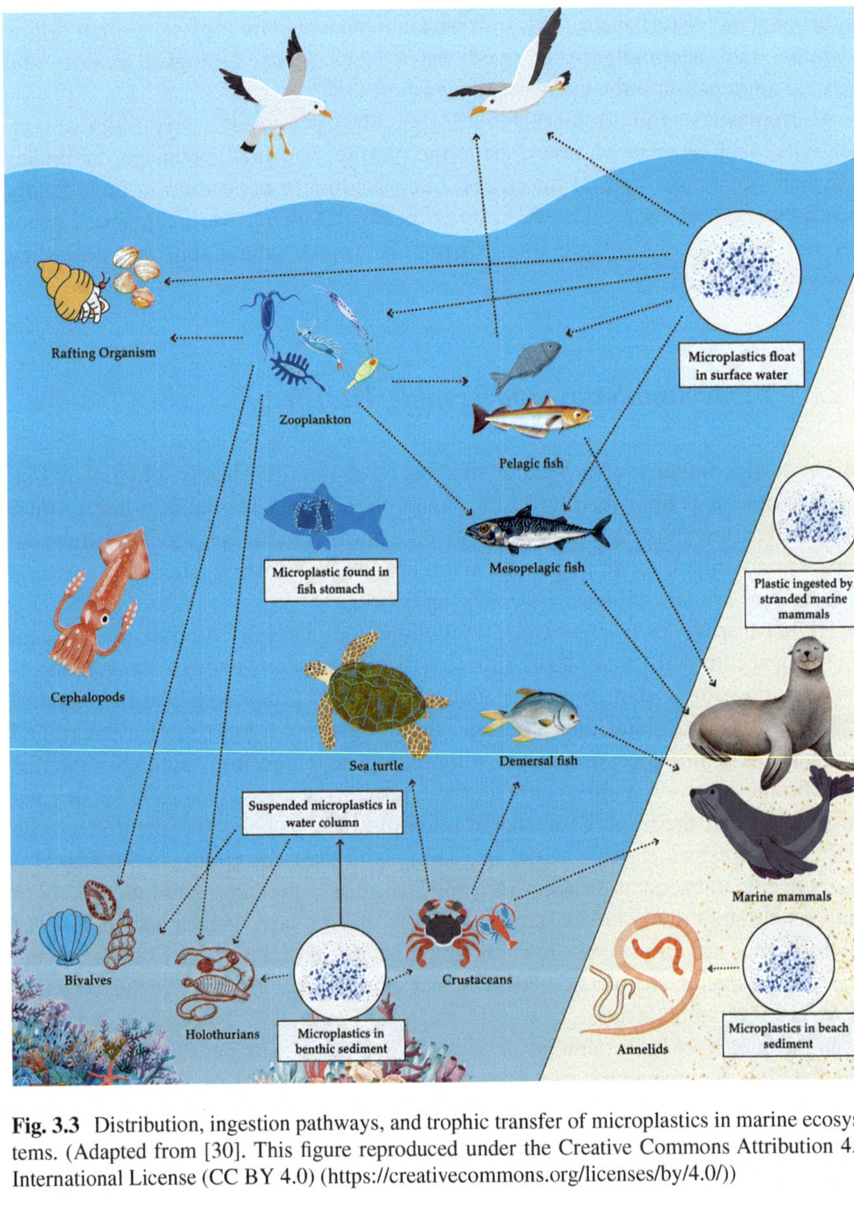

Fig. 3.3 Distribution, ingestion pathways, and trophic transfer of microplastics in marine ecosystems. (Adapted from [30]. This figure reproduced under the Creative Commons Attribution 4.0 International License (CC BY 4.0) (https://creativecommons.org/licenses/by/4.0/))

interact with a wide range of marine organisms. Primary exposure occurs through ingestion by zooplankton, rafting organisms, and benthic invertebrates, such as bivalves, holothurians, annelids, and crustaceans. These particles are subsequently transferred through trophic interactions to higher organisms, including cephalopods, pelagic and demersal fish, sea turtles, seabirds, and marine mammals. The figure also highlights the accumulation of microplastics in fish gastrointestinal tracts and the presence of plastic particles in beach sediments, demonstrating the connectivity between pelagic, benthic, and coastal environments. Together, the diagram emphasizes how microplastics move through marine ecosystems via feeding interactions, habitat exchange, and sediment interactions, illustrating their ecological persistence and potential implications for marine biodiversity and food web integrity.

Beyond their toxic effects, microplastics serve as vectors for hydrophobic organic pollutants (HOPs), heavy metals, and microbial pathogens. Their large surface-area-to-volume ratio enables efficient adsorption and concentration of contaminants from the surrounding media. These "contaminated carriers" can enhance the bioavailability and toxicological footprint of legacy pollutants, such as PCBs, PAHs, and endocrine-disrupting compounds [31, 32].

3.2.4 Soil Systems and Terrestrial Retention

Terrestrial environments receive microplastics through multiple vectors: atmospheric fallout, sewage sludge application, plastic mulch degradation, tire wear, and road runoff. Once in soils, microplastics persist due to low photodegradation and limited microbial enzymatic capability [33]. Their fate in soils depends on polymer type, shape, density, and interaction with organic matter and mineral particles.

Although vertical mobility in soils remains relatively slow, agricultural practices such as plowing or irrigation can enhance particle transport into deeper layers [34]. Microplastics can alter soil aggregation, porosity, water retention, and nutrient cycling. Root systems interact with microplastics through sorption and penetration, potentially influencing plant growth and microbiome stability. Recent findings reveal microplastic uptake into plant tissues, although quantification remains challenging [35].

Soil invertebrates, including earthworms and collembolans, inadvertently ingest microplastics, triggering physiological stress, reduced reproduction, and altered gut microbiota [36, 37]. These biological interactions influence microplastic redistribution and enhance bioavailability to higher trophic levels in terrestrial food webs. Figure 3.4 illustrates the major sources, transformation mechanisms, and environmental fate of microplastics within agricultural soil systems. Key entry pathways include the degradation of plastic mulching films, application of sewage sludge and compost, wastewater irrigation, landfill leachate, improper plastic disposal, and atmospheric deposition. Once introduced into soils, microplastics undergo physical fragmentation, chemical reactions, and biological processes that influence their

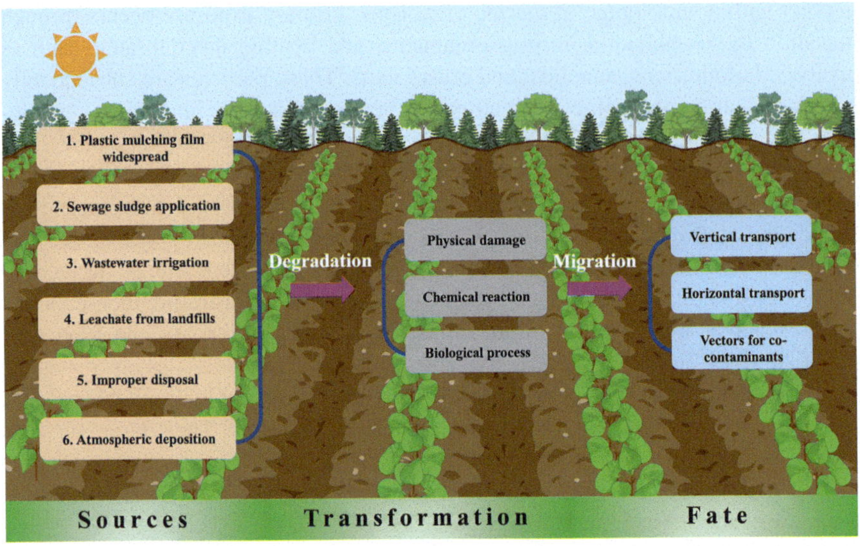

Fig. 3.4 Sources, transformation processes, and environmental fate of microplastics in agricultural soils. (Adapted from [38]. This figure reproduced under the Creative Commons Attribution 4.0 International License (CC BY 4.0) (https://creativecommons.org/licenses/by/4.0/))

persistence and interactions with soil matrices. The figure further highlights the environmental fate of these particles, including vertical migration into deeper soil layers, horizontal transport through surface and subsurface flows, and their role as vectors for co-contaminants, such as heavy metals and organic pollutants. Together, the diagram emphasizes how agricultural practices, environmental processes, and soil dynamics collectively determine the distribution, mobility, and ecological risks of microplastics in terrestrial ecosystems.

3.2.5 Atmospheric Transport and Global Dispersion

Microplastics are increasingly recognized as airborne contaminants originating from urban emissions, wind resuspension, and road dust [9, 21]. Fibers and fragments enter the atmosphere through multiple anthropogenic activities, including textile abrasion, tire wear, industrial discharges, recycling operations, and agricultural practices. Once suspended in the atmosphere, these particles undergo turbulent mixing and are transported by atmospheric circulation, enabling them to move across regional and intercontinental scales. This long-range transport has been confirmed by the detection of microplastics in remote environments, including mountainous regions, deep ocean areas, and polar ecosystems. Figure 3.5 contextualizes these processes by presenting the global atmospheric budget of microplastics and highlighting the atmosphere as a major pathway for their redistribution. The figure

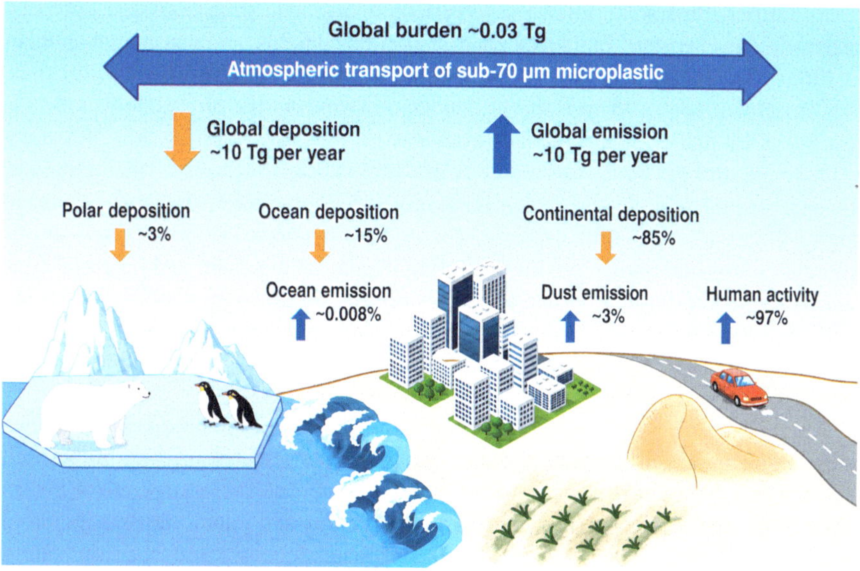

Fig. 3.5 Global atmospheric emissions, transport, and deposition pathways of airborne microplastics. (Adapted from [39]. This figure reproduced under the Creative Commons Attribution 4.0 International License (CC BY 4.0) (https://creativecommons.org/licenses/by/4.0/))

indicates that global emissions and deposition are approximately balanced at about 10 Tg per year, while the atmospheric burden remains relatively small (~0.03 Tg), reflecting the dynamic balance between emission, transport, and removal processes. It further shows that most atmospheric emissions (~97%) originate from human activities, particularly those associated with urbanization, transportation, and industrial production, whereas natural dust sources contribute a smaller fraction (~3%).

This figure presents the global atmospheric budget and transport dynamics of microplastics smaller than 70 μm, highlighting the balance between global emissions (~10 Tg per year), atmospheric burden (~0.03 Tg), and annual deposition fluxes (~10 Tg per year). The distribution patterns indicate that continental regions receive the largest share of atmospheric deposition (~85%), followed by oceanic (~15%) and polar (~3%) environments, demonstrating the ability of atmospheric circulation to redistribute microplastics across both densely populated and remote ecosystems. Anthropogenic activities account for the dominant share of emissions (~97%), largely associated with urbanization, transportation, and industrial processes, while natural dust emissions contribute a comparatively small fraction (~3%). Although the oceans receive substantial atmospheric inputs, their contribution as emission sources remains negligible (~0.008%), suggesting that marine environments function primarily as sinks rather than significant atmospheric sources. Together, these patterns emphasize the atmosphere as a major redistribution pathway within the global microplastic cycle [39].

As further illustrated, atmospheric pathways originate from emission hotspots, such as traffic corridors, industrial zones, agricultural fields, and waste management facilities, which release airborne microplastics that are transported by wind from local to global scales. Atmospheric deposition connects the atmosphere with terrestrial and aquatic environments through wet processes, such as rain and snow scavenging, and dry processes, such as gravitational settling and surface interception. Deposited particles may also be resuspended, creating a dynamic exchange among environmental compartments. These processes have important implications for ecosystems and human health, including inhalation exposure and potential contamination of food systems. The figure, therefore, highlights the atmosphere as an active interface for exchange among terrestrial, aquatic, and biological systems and underscores the need to integrate atmospheric transport into microplastic monitoring, exposure assessment, and pollution control strategies.

Atmospheric microplastics exhibit high deposition fluxes in urban environments and coastal areas. Precipitation events scavenge particles from the atmosphere, transferring them into terrestrial and aquatic systems. These deposition cycles expand the environmental footprint of microplastics and enable cross-ecosystem transport [10]. Re-entrainment from surfaces creates feedback loops, complicating source attribution and exposure mapping.

Studies report human inhalation exposure through indoor air and occupational settings, with potential links to respiratory irritation, oxidative stress, and inflammation. The health implications of long-term inhalation remain largely unknown, particularly for small-sized fractions, such as nanoplastics and volatile plastic additives [40].

3.2.6 From Particles to Pollutant Complexes

The fate of microplastics depends not only on physical properties, such as size, shape, and density, but also on their chemical composition and environmental aging. Weathered microplastics exhibit altered surface chemistry, often with increased roughness and oxidation that enhance reactivity and sorption capacity. This aging process transforms microplastics into complex contaminant matrices capable of interacting with organic matter, metal ions, and microbial consortia [41].

Plastic additives such as flame retardants, stabilizers, plasticizers, and pigments leach into the environment, contributing to ecotoxicological effects. These leachates may induce hormonal disruption, genotoxicity, and immune suppression in exposed organisms [42]. In turn, adsorbed pollutants amplify chemical risks, transforming microplastics from passive debris to active pollutant complexes.

Emerging research on nanoplastics—a degradation product of microplastics—reveals even greater ecological and cellular impacts. Their small size facilitates translocation across biological membranes, including the blood-brain barrier, and accumulation in tissues [43]. Nanoplastics exhibit enhanced reactivity and surface area, magnifying their role as vectors and cellular disruptors in aquatic and terrestrial biota.

3.3 Case Studies: Global and Southeast Asian Hotspots of Microplastic Pollution

Microplastic contamination does not distribute evenly across the planet. Instead, it converges in specific ecological and urban regions characterized by intense industrial activity, rapid population growth, and ineffective waste management. These hotspots, both in marine and freshwater systems, constitute critical zones of concern for policymakers, researchers, and environmental managers. Understanding the dynamics behind their formation is crucial for designing interventions tailored to local and regional contexts.

3.3.1 Global Accumulation Zones

In the world's oceans, floating plastics concentrate within major oceanic gyres due to surface current circulation. The North Pacific Subtropical Gyre, colloquially termed the Great Pacific Garbage Patch, exemplifies this process. This region holds an estimated 1.8 trillion pieces of plastic, many in the microplastic size range, distributed across a surface area twice the size of Texas [44]. Such floating aggregations trap particles from Asia and North America and pose ingestion risks to pelagic species.

Europe's Danube River is a significant source of microplastics entering the Black Sea. Research estimates suggest that the Danube delivers up to 4.2 tonnes of plastic debris daily, with microplastics accounting for a substantial fraction [45]. The majority originates from mismanaged urban waste, industrial runoff, and combined sewer overflows.

The Yangtze River in China, often cited as the single largest contributor of riverborne plastic to the marine environment, reflects the intersection of dense urban populations and manufacturing hubs with limited enforcement of waste-disposal regulations [46]. The river discharges millions of microplastic particles per cubic meter of water, which eventually reach the East China Sea.

3.3.1.1 Global Microplastic Hotspot

- North Pacific Gyre (USA-Japan)
  Marine convergence zone exceeding 1.8 trillion plastic particles, dominated by low-density polyethylene and polypropylene fragments [44].

- Danube River (Europe)
  Delivers over 4 tonnes of plastics per day, heavily influenced by untreated urban wastewater [45].
- Yangtze River (China)

  Microplastic densities exceeding 10,000 particles per cubic meter in urban river sections [46].

3.3.2 Case Studies in Europe

In Europe, extensive monitoring efforts have revealed widespread microplastic presence in rivers, estuaries, and sediments. The Thames River in the UK records high concentrations of microbeads and synthetic fibers, especially near wastewater discharge points [47]. The Rhine River, shared by six countries, also shows high levels of microplastics due to industrial activities and dense population clusters [48].

Mediterranean beaches and coastal sediments accumulate microplastics at alarming densities. Seasonal tourism and maritime traffic exacerbate plastic leakage into the sea [49]. Despite advanced waste treatment systems, fragmentation of legacy plastics and atmospheric fallout continue to pollute the environment.

3.3.3 Case Studies in Asia (Southeast Asian Case Focus)

Southeast Asia stands at the intersection of global plastic production, rapid urbanization, and extensive coastal development. These conditions make the region particularly vulnerable to microplastic pollution. Numerous studies confirm that Southeast Asia contributes significantly to global marine plastic discharge, with rivers, estuaries, and coastal megacities acting as conduits for plastic debris into marine ecosystems [6, 50]. The region's climatic factors, economic activities, and fragmented waste governance amplify the challenge of controlling microplastic dispersal and accumulation.

3.3.3.1 Rivers as Transport Corridors

Major rivers in Southeast Asia play a pivotal role in delivering land-based plastics to the ocean. Among the most studied is the Mekong River, which originates in China and flows through Myanmar, Laos, Thailand, Cambodia, and Vietnam. It serves as a transboundary plastic highway, discharging substantial microplastic loads into the South China Sea. Thành et al. [51] reported that plastic debris from urban areas in Phnom Penh and Ho Chi Minh City enters tributaries that merge with the Mekong, facilitating downstream transport. During monsoonal floods, retention basins and floodplains act as temporary reservoirs, eventually flushing accumulated particles into estuarine systems.

Similarly, Thailand's Chao Phraya River, which drains the greater Bangkok area, has shown consistent patterns of microplastic contamination along its course. Babel et al. [52] observed polymer fragments and fibers in surface waters, linked to household discharges, textile dyeing operations, and urban runoff. Seasonal variations, such as peak flow during rainy months, elevate the volume and transport distance of microplastics. In Indonesia, the Citarum River, often labeled as one of the world's most polluted waterways, presents a convergence of industrial effluent, informal settlements, and agricultural plastic use. Studies in West Java have revealed polyethylene, polypropylene, and polyester particles in sediment and water samples, with upstream-to-downstream accumulation gradients [53].

3.3.3.2 Urban Estuaries and Coastal Sediments

Southeast Asian coastal cities have become critical accumulation zones for microplastics. In Manila Bay, the Philippines, surface waters and sediments exhibit high concentrations of microplastics, dominated by fibers, fragments, and films [54]. Tanchuling and Osorio [55] linked fiber dominance to wastewater discharge from the city, including effluents from laundry, textile processing, and domestic graywater. Tidal mixing, limited wastewater treatment infrastructure, and unregulated dumpsites exacerbate the problem. Surface water samples from Sitio Pulo, Navotas, revealed abundant microplastic fragments dominated by polyethylene (PE) and polypropylene (PP), exhibiting rounded shapes and discoloration, which indicate ongoing weathering and persistent plastic contamination in the aquatic environment [56].

In Indonesia, Jakarta Bay faces similar challenges. The bay receives untreated waste from over 13 rivers crossing the capital [57]. Dwiyitno et al. [58] identified polypropylene and polystyrene particles, likely derived from packaging materials, food containers, and industrial discharges. Mangrove zones in the bay's fringes act as physical traps for microplastics, with seagrass meadows also showing entrapment potential. These findings underscore the importance of vegetated coastal habitats in modulating microplastic distribution but also raise concerns over trophic transfer within these biodiverse ecosystems.

3.3.3.3 Blue Carbon Ecosystems as Pollution Sinks

Mangrove forests and seagrass beds, widely distributed along the coasts of Malaysia, Indonesia, the Philippines, and Thailand, serve as microplastic filters through their dense root systems and sediment-trapping capacity. Tan and Mohd Zanuri [59] reported the abundance and distribution of microplastics in tropical estuarine mangrove areas around Penang, Malaysia, documenting their presence across multiple sites within these mangrove ecosystems. Fragments and fibers were the dominant microplastic forms recorded in the mangrove areas around Penang, suggesting inputs from degraded consumer plastics and textile-related sources. Their accumulation in estuarine waters and sediments highlights potential ecological risks to mangrove-associated organisms, including benthic invertebrates and microbial communities that support coastal food webs.

Similarly, in the southern Philippines, microplastics have been documented in farmed seaweed (*Kappaphycus alvarezii*) and surrounding waters in Panguil Bay, as well as in estuarine sediments and edible bivalves, demonstrating their occurrence across aquaculture systems and benthic compartments of coastal environments [60, 61]. These findings raise concerns over the incorporation of microplastics into coastal food webs, particularly through aquaculture species such as farmed *Kappaphycus alvarezii* and filter-feeding bivalves that can ingest suspended particles. The documented presence of microplastics in sediments and edible mollusks further suggests potential ecological and human health implications, as particle accumulation in benthic environments may influence sediment quality and trophic transfer within estuarine systems.

3.3.3.4 Human Exposure and Food Chain Integration

Southeast Asian populations heavily rely on seafood, and evidence suggests that microplastics have entered regional food chains. Mussels and oysters collected from aquaculture farms in Vietnam and the Philippines exhibit microplastic contamination in their digestive tissues [61, 62]. In Thailand, microplastic contamination was documented in edible marine fishes from the upper Gulf of Thailand, with nearly half of the sampled individuals containing microplastics, and higher contamination observed in demersal species than pelagic ones. The predominance of fibrous and polyester particles highlights potential exposure through feeding habitats, underscoring the need for continued evaluation of dietary exposure and associated health risks [63].

3.3.3.5 Governance and Monitoring Challenges

Despite mounting evidence of widespread contamination, most Southeast Asian nations lack standardized protocols for monitoring microplastics. Research efforts remain concentrated in academic institutions and are often fragmented. Regional databases are rare, and national marine debris action plans often prioritize macroplastics. While ASEAN has elevated marine plastic waste management through collective declarations and regional initiatives, progress in deepening implementation across member states remains constrained by disparities in funding, technological capacity, and management practices [64]. Singapore and Malaysia have begun limited source-tracking and beach-monitoring programs, but comprehensive legislation addressing microplastic risks is still in development.

Compounding the issue is the lack of laboratory infrastructure capable of detecting nanoplastics or characterizing complex microplastic mixtures. The need for harmonized methodologies spanning sampling, extraction, identification, and quantification is evident across published case studies. Without shared standards, comparing results and formulating policy responses remains difficult.

3.3.3.6 Southeast Asia's Microplastic Profile

- **Rivers as conduits**: Mekong, Chao Phraya, and Citarum Rivers transport microplastics into key marine zones.
- **Urban hotbeds**: Manila Bay and Jakarta Bay sediments show high microplastic accumulation.
- **Blue carbon threats**: Mangroves and seagrasses trap microplastics, with risks to fisheries and biodiversity.
- **Weak governance**: Monitoring remains fragmented, and microplastic-specific policy responses lag behind regional needs.

3.4 Chapter Synthesis

This chapter establishes that microplastic pollution operates as a dynamic environmental process rather than a static contamination problem. Evidence presented across terrestrial, freshwater, marine, and atmospheric systems demonstrates that microplastics follow predictable transport pathways shaped by hydrology, meteorology, land use, and human activity patterns. Rivers function as major conduits linking inland sources to coastal environments. Atmospheric transport expands the geographic reach of microplastics beyond immediate emission zones. Soil systems act both as sinks and secondary sources through resuspension and runoff processes. These observations confirm that microplastic distribution reflects environmental connectivity.

The discussion on environmental fate further shows that persistence defines the long-term challenge of microplastic pollution. Once deposited in sediments, soils, or biological systems, particles may remain for extended periods while undergoing slow fragmentation and chemical weathering. Biofouling, aggregation, and interaction with organic contaminants modify their environmental behavior. These processes complicate predictions of transport and ecological exposure. Long residence times increase the probability of biological interaction and trophic transfer. Environmental accumulation, therefore, depends on both emission rates and degradation dynamics.

Regional case studies reinforce the uneven global distribution of microplastic pollution. Rapid urbanization, high coastal population density, riverine discharge, and limited waste infrastructure contribute to high environmental loads in Southeast Asia. These regional examples highlight how socioeconomic conditions influence environmental exposure. They also demonstrate the importance of localized monitoring data in understanding pollution patterns. This chapter, therefore, emphasizes that global assessments must integrate regional realities to support effective mitigation strategies.

Taken together, the evidence presented in this chapter shows that microplastic pollution reflects the interaction between material life cycles and environmental processes. Addressing this challenge requires scientific understanding of transport mechanisms, environmental persistence, and regional variability. Such knowledge provides the basis for monitoring programs, risk assessment, and technological intervention.

Understanding environmental pathways explains where microplastics accumulate and why they are exposed to. Scientific progress now depends on the ability to detect, quantify, and characterize these particles with accuracy and consistency. The next chapter, therefore, examines the analytical foundations of microplastic research. It explores the tools used to identify microplastics and nanoplastics, the methodological challenges that influence data reliability, and emerging innovations in monitoring technologies. If this chapter explains where microplastics travel and persist, the next chapter explains how science makes them visible and measurable.

References

1. Hahladakis, J. N., Velis, C. A., Weber, R., Iacovidou, E., & Purnell, P. (2022). An overview of chemical additives present in plastics: Migration, release, fate and environmental impact during their use, disposal and recycling. *Journal of Hazardous Materials, 344*, 179–199. https://doi.org/10.1016/j.jhazmat.2017.10.014
2. Geyer, R., Jambeck, J. R., & Law, K. L. (2017). Production, use, and fate of all plastics ever made. *Science Advances, 3*(7), e1700782. https://doi.org/10.1126/sciadv.1700782
3. Tran, T. K. A., Raju, S., Singh, A., Senathirajah, K., Bhagwat-Russell, G., Daggubati, L., Kandaiah, R., & Palanisami, T. (2023). Occurrence and distribution of microplastics in long-term biosolid-applied rehabilitation land: An overlooked pathway for microplastic entry into terrestrial ecosystems in Australia. *Environmental Pollution, 336*, 122464. https://doi.org/10.1016/j.envpol.2023.122464
4. Lusher, A., Hollman, P., & Mendoza-Hill, J. (2015). Microplastics in fisheries and aquaculture: Status of knowledge on their occurrence and implications for aquatic organisms and food safety. *FAO Fisheries and Aquaculture Technical Paper*, No. 615.
5. Horton, A. A., Walton, A., Spurgeon, D. J., Lahive, E., & Svendsen, C. (2017). Microplastics in freshwater and terrestrial environments: Evaluating the current understanding to identify the knowledge gaps and future research priorities. *Science of the Total Environment, 586*, 127–141. https://doi.org/10.1016/j.scitotenv.2017.01.190
6. Lebreton, L. C. M., van der Zwet, J., Damsteeg, J. W., Slat, B., Andrady, A., & Reisser, J. (2017). River plastic emissions to the world's oceans. *Nature Communications, 8*(1), 15611. https://doi.org/10.1038/ncomms15611
7. Chae, Y., & An, Y. J. (2018). Current research trends on plastic pollution and ecological impacts on the soil ecosystem: A review. *Environmental Pollution, 240*, 387–395. https://doi.org/10.1016/j.envpol.2018.05.008

8. Wright, S. L., Thompson, R. C., & Galloway, T. S. (2013). The physical impacts of microplastics on marine organisms: A review. *Environmental Pollution, 178*, 483–492. https://doi.org/10.1016/j.envpol.2013.02.031
9. Dris, R., Gasperi, J., Saad, M., Mirande, C., & Tassin, B. (2016). Synthetic fibers in atmospheric fallout: A source of microplastics in the environment? *Marine Pollution Bulletin, 104*(1–2), 290–293. https://doi.org/10.1016/j.marpolbul.2016.01.006
10. Brahney, J., Hallerud, M., Heim, E., Hahnenberger, M., & Sukumaran, S. (2020). Plastic rain in protected areas of the United States. *Science, 368*(6496), 1257–1260. https://doi.org/10.1126/science.aaz5819
11. Rochman, C. M., & Hoellein, T. (2020). The global odyssey of plastic pollution. *Science, 368*(6496), 1184–1185. https://doi.org/10.1126/science.abc4428
12. Bao, M., Xiang, X., Huang, J., Kong, L., Wu, J., & Cheng, S. (2023). Microplastics in the atmosphere and water bodies of coastal agglomerations: A mini-review. *International Journal of Environmental Research and Public Health, 20*(3), 2466. https://doi.org/10.3390/ijerph20032466
13. Praveena, S. M. (2025). Sources and occurrence of microplastics in the terrestrial environment. In N. Kataria, V. K. Garg, C. H. Han, & E. R. Rane (Eds.), *Microplastic pollution: Occurrence, health risk and challenges* (1st ed., pp. 68–84). CRC Press. https://doi.org/10.1201/9781032706573-5. ISBN: 9781032706573.
14. Tiwari, E., & Sistla, S. A. (2024). Agricultural plastic pollution reduces soil function even under best management practices. *PNAS Nexus, 3*(10), pgae433. https://doi.org/10.1093/pnasnexus/pgae433
15. Fei, X., He, H., Pi, X., Lu, X., Chen, Q., Ma, J., Wang, Y., Fang, B., Wu, C. X., & Feng, S. (2022). The distribution, behavior, and release of macro- and micro-size plastic wastes in solid waste disposal sites. *Critical Reviews in Environmental Science and Technology, 53*(3), 366–389. https://doi.org/10.1080/10643389.2022.2054649
16. McNulty, K., & Nahar, K. (2024). Geospatial modelling predicts agricultural microplastic hotspots from biosolid application risks. *Agronomy, 15*(1), 47. https://doi.org/10.3390/agronomy15010047
17. Fan, Y., Zheng, J., Xu, W., Zhang, Q., Chen, N., Wang, H., Qian, X., & Wang, G. (2024). Spatiotemporal occurrence and characteristics of microplastics in the urban road dust in a megacity, eastern China. *Journal of Hazardous Materials, 468*, 133733. https://doi.org/10.1016/j.jhazmat.2024.133733
18. Eerkes-Medrano, D., Thompson, R. C., & Aldridge, D. C. (2015). Microplastics in freshwater systems: A review of the emerging threats, identification of knowledge gaps and future research needs. *Water Research, 75*, 63–82. https://doi.org/10.1016/j.watres.2015.02.012
19. Na, S.-H., Kim, M.-J., Kim, J., Batool, R., Cho, K., Chung, J., Lee, S., & Kim, E. (2023). Fate and potential risks of microplastic fibers and fragments in water and wastewater treatment processes. *Journal of Hazardous Materials, 463*, 132938. https://doi.org/10.1016/j.jhazmat.2023.132938
20. Cózar, A., Echevarría, F., González-Gordillo, J. I., Irigoien, X., Úbeda, B., Hernández-León, S., Palma, A. T., Navarro, S., García-de-Lomas, J., Ruiz, A., Fernández-de-Puelles, M. L., & Duarte, C. M. (2014). Plastic debris in the open ocean. *Proceedings of the National Academy of Sciences of the United States of America, 111*(28), 10239–10244. https://doi.org/10.1073/PNAS.1314705111
21. Allen, S., Allen, D., Phoenix, V. R., Le Roux, G., Durantez Jimenez, P., Simonneau, A., Binet, S., & Galop, D. (2019). Atmospheric transport and deposition of microplastics in a remote mountain catchment. *Nature Geoscience, 12*(5), 339–344. https://doi.org/10.1038/S41561-019-0335-5
22. Andrady, A. L. (2011). Microplastics in the marine environment. *Marine Pollution Bulletin, 62*(8), 1596–1605. https://doi.org/10.1016/j.marpolbul.2011.05.030
23. GESAMP. (2015). Sources, fate and effects of microplastics in the marine environment: A global assessment. In *IMO/FAO/UNESCO-IOC/UNIDO/WMO/IAEA/UN/UNEP/UNDP Joint Group of Experts on the Scientific Aspects of Marine Environmental Protection.*

24. Arif, Y., Mir, A. R., Zieliński, P., Hayat, S., & Bajguz, A. (2024). Microplastics and nanoplastics: Source, behavior, remediation, and multi-level environmental impact. *Journal of Environmental Management, 356*, 120618. https://doi.org/10.1016/j.jenvman.2024.120618
25. Rynek, R., Tekman, M. B., Rummel, C., Bergmann, M., Wagner, S., Jahnke, A., & Reemtsma, T. (2024). Hotspots of floating plastic particles across the North Pacific Ocean. *Environmental Science & Technology, 58*(9), 4302–4313. https://doi.org/10.1021/acs.est.3c05039
26. Corti, A., La Nasa, J., Biale, G., Ceccarini, A., Manariti, A., Modugno, F., & Castelvetro, V. (2023). Microplastic pollution in the sediments of interconnected lakebed, seabed, and seashore aquatic environments: Polymer-specific total mass through the multianalytical "PISA" procedure. *Analytical and Bioanalytical Chemistry, 415*, 2921–2936. https://doi.org/10.1007/s00216-023-04664-0
27. Blettler, M. C. M., Ulla, M. A. d. H., Rabuffetti, A. P., & Garello, N. (2017). Plastic pollution in freshwater ecosystems: Macro-, meso-, and microplastic debris in a floodplain lake. *Environmental Monitoring and Assessment, 189*(11), 581. https://doi.org/10.1007/S10661-017-6305-8
28. Barboza, L. G. A., Vieira, L. R., Branco, V., Figueiredo, N., Carvalho, F., Carvalho, C., & Guilhermino, L. (2018). Microplastics cause neurotoxicity, oxidative damage and energy-related changes and interact with the bioaccumulation of mercury in the European seabass, Dicentrarchus labrax (Linnaeus, 1758). *Aquatic Toxicology, 195*, 49–57. https://doi.org/10.1016/j.aquatox.2017.12.008
29. Li, X., Wei, Y., Wang, F., Liu, X., Su, H., & Zhang, R. (2023). Occurrence, bioaccumulation, and risk assessment of microplastics in the aquatic environment: A review. *Water, 15*(9), 1768. https://doi.org/10.3390/w15091768
30. Sunny, A. R., Sazzad, S. A., Islam, M. A., Mithun, M. H., Hussain, M., Raposo, A., & Bhuiyan, M. K. A. (2025). Microplastics in aquatic ecosystems: A global review of distribution, ecotoxicological impacts, and human health risks. *Water, 17*(12), 1741. https://doi.org/10.3390/w17121741
31. Weis, J. S., & Alava, J. J. (2023). (Micro)plastics are toxic pollutants. *Toxics, 11*, 935. https://doi.org/10.3390/toxics11110935
32. Brennecke, D., Duarte, B., Paiva, F., Caçador, I., & Canning-Clode, J. (2016). Microplastics as vector for heavy metal contamination from the marine environment. *Estuarine, Coastal and Shelf Science, 178*, 189–195. https://doi.org/10.1016/j.ecss.2015.12.003
33. He, D., Luo, Y., Lu, S., Liu, M., Song, Y., & Lei, L. (2018). Microplastics in soils: Analytical methods, pollution characteristics and ecological risks. *Trends in Analytical Chemistry, 109*, 163–172. https://doi.org/10.1016/J.TRAC.2018.10.006
34. Veenstra, J. J., & Burras, C. L. (2015). Soil profile transformation after 50 years of agricultural land use. *Soil Science Society of America Journal, 79*(4), 1154–1162. https://doi.org/10.2136/SSSAJ2015.01.0027
35. Zhai, Y., Bai, J., Chang, P., Liu, Z., Wang, Y., Liu, G., Cui, B., Peijnenburg, W., & Vijver, M. G. (2024). Microplastics in terrestrial ecosystem: Exploring the menace to the soil-plant-microbe interactions. *TrAC Trends in Analytical Chemistry, 174*, 117667. https://doi.org/10.1016/j.trac.2024.117667
36. Zhu, D., Chen, Q.-L., An, X.-L., Yang, X.-R., Christie, P., Ke, X., Wu, L., & Zhu, Y.-G. (2018). Exposure of soil collembolans to microplastics perturbs their gut microbiota and alters their isotopic composition. *Soil Biology & Biochemistry, 116*, 302–310. https://doi.org/10.1016/J.SOILBIO.2017.10.027
37. Chan, W. T., Medriano, C. A., & Bae, S. (2023). Unveiling the impact of short-term polyethylene microplastics exposure on metabolomics and gut microbiota in earthworms (*Eudrilus eugeniae*). *Journal of Hazardous Materials, 460*, 132305. https://doi.org/10.1016/j.jhazmat.2023.132305
38. Chen, Z., Carter, L. J., Banwart, S. A., & Kay, P. (2025). Microplastics in soil–plant systems: Current knowledge, research gaps, and future directions for agricultural sustainability. *Agronomy, 15*(7), 1519. https://doi.org/10.3390/agronomy15071519

39. Yang, S., Brasseur, G. P., Walters, S., Lichtig, P., & Li, C. W. Y. (2025). Global atmospheric distribution of microplastics with evidence of low oceanic emissions. *NPJ Climate and Atmospheric Science, 8*, 81. https://doi.org/10.1038/s41612-025-00914-3
40. Bhardwaj, G., Abdulkadhim, M. H., Joshi, K., Wankhede, L., Das, R., & Brar, S. K. (2025). Exposure pathways, systemic distribution, and health implications of micro- and nanoplastics in humans. *Applied Sciences, 15*(16), 8813. https://doi.org/10.3390/app15168813
41. Binda, G., Kalčíková, G., Allan, I. J., Hurley, R., Rødland, E., Spanu, D., & Nizzetto, L. (2024). Microplastic aging processes: Environmental relevance and analytical implications. *TrAC Trends in Analytical Chemistry, 172*, 117566. https://doi.org/10.1016/j.trac.2024.117566
42. Vilke, J. M., Fonseca, T. G., Alkimin, G. D., Gonçalves, J. M., Edo, C., d' Errico, G., Seilitz, F. S., Rotander, A., Benedetti, M., Regoli, F., Lüchmann, K. H., & Bebianno, M. J. (2024). Looking beyond the obvious: The ecotoxicological impact of the leachate from fishing nets and cables in the marine mussel Mytilus galloprovincialis. *Journal of Hazardous Materials, 473*, 134479. https://doi.org/10.1016/j.jhazmat.2024.134479
43. Gigault, J., ter Halle, A., Baudrimont, M., Pascal, P.-Y., Gauffre, F., Phi, T.-L., El Hadri, H., Grassl, B., & Reynaud, S. (2018). Current opinion: What is a nanoplastic? *Environmental Pollution, 235*, 1030–1034. https://doi.org/10.1016/j.envpol.2018.01.024
44. Lebreton, L., Slat, B., Ferrari, F., Sainte-Rose, B., Aitken, J., Marthouse, R., Hajbane, S., Cunsolo, S., Schwarz, A., Levivier, A., Noble, K., Debeljak, P., Maral, H., Schoeneich-Argent, R., Brambini, R., & Reisser, J. (2018). Evidence that the Great Pacific Garbage Patch is rapidly accumulating plastic. *Scientific Reports, 8*, 4666. https://doi.org/10.1038/s41598-018-22939-w
45. Lechner, A., Keckeis, H., Lumesberger-Loisl, F., Zens, B., Krusch, R., Tritthart, M., Glas, M., & Schludermann, E. (2014). The Danube so colourful: A potpourri of plastic litter outnumbers fish larvae in Europe's second largest river. *Environmental Pollution, 188*, 177–181. https://doi.org/10.1016/j.envpol.2014.02.006
46. Zhao, S., Zhu, L., & Li, D. (2015). Microplastic in three urban estuaries, China. *Environmental Pollution, 206*, 597–604. https://doi.org/10.1016/j.envpol.2015.08.027
47. West-Clarke, Z., & Turner, A. (2024). Contamination of Thames Estuary sediments by retroreflective glass microbeads, road marking paint fragments and anthropogenic microfibres. *Science of the Total Environment, 912*, 169257. https://doi.org/10.1016/j.scitotenv.2023.169257
48. Mani, T., Hauk, A., Walter, U., & Burkhardt-Holm, P. (2016). Microplastics profile along the Rhine River. *Scientific Reports, 5*, 17988. https://doi.org/10.1038/srep17988
49. Soursou, V., Campo, J., & Picó, Y. (2024). Spatio-temporal variation and ecological risk assessment of microplastics along the touristic beaches of a Mediterranean coast transect (Valencia province, East Spain). *Journal of Environmental Management, 354*, 120315. https://doi.org/10.1016/j.jenvman.2024.120315
50. Jambeck, J. R., Geyer, R., Wilcox, C., Siegler, T. R., Perryman, M., Andrady, A., Narayan, R., & Law, K. L. (2015). Plastic waste inputs from land into the ocean. *Science, 347*(6223), 768–771. https://doi.org/10.1126/science.1260352
51. Thành, N. T., Minh, H. V. T., Lavane, K., Ngan, N. V. C., Toan, P. V., Ty, T. V., Duy, D. V., Toan, V. T., & Kumar, P. (2025). From river to sea: Tracking plastic waste transport via the Hau River, Mekong Delta, Vietnam. *Water, 17*(16), 2438. https://doi.org/10.3390/w17162438
52. Babel, S., Ta, A. T., Nguyen, T. P. L., Sembiring, E., Setiadi, T., & Sharp, A. (2022). Microplastics pollution in selected rivers from Southeast Asia. *APN Science Bulletin, 12*(1), 5–17. https://doi.org/10.30852/sb.2022.1741
53. Sembiring, E., Fareza, A. A., Suendo, V., & Reza, M. (2020). The presence of microplastics in water, sediment, and milkfish (*Chanos chanos*) at the downstream area of Citarum River, Indonesia. *Water, Air, & Soil Pollution, 231*, 355. https://doi.org/10.1007/s11270-020-04757-5
54. Osorio, E. D., Tanchuling, M. A. N., & Diola, M. B. L. D. (2021). Microplastics occurrence in surface waters and sediments in five river mouths of Manila Bay. *Frontiers in Environmental Science, 9*, 719274. https://doi.org/10.3389/fenvs.2021.719274
55. Tanchuling, M. A. N., & Osorio, E. D. (2020). The microplastics in Metro Manila rivers: Characteristics, sources, and abatement. In F. Stock, G. Reifferscheid, N. Brennholt, &

E. Kostianaia (Eds.), *Plastics in the aquatic environment—Part I* (pp. 405–426). Springer. https://doi.org/10.1007/698_2020_543

56. Galicia, M. L. C., Festado, C. L. P., Magalong, J. R. S., Rivera, R. B. L., Villacorte, K. M., Sarmiento, W. P. G., & Deocaris, C. C. (2023). Assessment of microplastics in the surface water of Sitio Pulo, Navotas, Metro Manila. *Applied Environmental Research, 45*(1), 1. https://doi.org/10.35762/AER.2023001
57. Diastomo, H., Surya, M. Y., Sakti, A. D., Agustina, E., & Trismadi. (2021). Marine debris tracking from river discharge base on hydrodynamic simulation on Jakarta Bay. *IOP Conference Series: Earth and Environmental Science, 925*, 012005. https://doi.org/10.1088/1755-1315/925/1/012005
58. Dwiyitno, D., Andayani, F., & Wibowo, S. (2020). Concentration and characteristic of floating plastic debris in Jakarta Bay: A preliminary study. *Squalen Bulletin of Marine and Fisheries Postharvest and Biotechnology, 15*(3), 462–469. https://doi.org/10.15578/squalen.v15i3.462
59. Tan, E., & Mohd Zanuri, N. B. (2023). Abundance and distribution of microplastics in tropical estuarine mangrove areas around Penang, Malaysia. *Frontiers in Marine Science, 10*, 1148804. https://doi.org/10.3389/fmars.2023.1148804
60. Reble, D., Capangpangan, R. Y., Espadero, A. D. A., Leopardas, V. E., & Uy, W. H. (2025). The prevalence of microplastics in farmed seaweed *Kappaphycus alvarezii* in Panguil Bay, Philippines. *IOP Conference Series: Earth and Environmental Science, 1506*, 012028. https://doi.org/10.1088/1755-1315/1506/1/012028
61. Bonifacio, P. S. P., Metillo, E. B., & Romano, E. F., Jr. (2022). Microplastic in sediments and ingestion rates in three edible bivalve Mollusc species in a southern Philippine estuary. *Water, Air, & Soil Pollution, 233*, 455. https://doi.org/10.1007/s11270-022-05928-4
62. Dao, T.-S., Kieu-Le, T.-C., La, X.-T., Truong, N.-T.-S., Thuong, Q.-T., Nguyen, V.-T., Nguyen, Q.-H., & Strady, E. (2023). Microplastic accumulation in oysters: Insights from aquaculture and laboratory conditions. *Regional Studies in Marine Science, 68*, 103251. https://doi.org/10.1016/j.rsma.2023.103251
63. Srisiri, S., Haetrakul, T., Dunbar, S. G., & Chansue, N. (2024). Microplastic contamination in edible marine fishes from the upper Gulf of Thailand. *Marine Pollution Bulletin, 198*, 115785. https://doi.org/10.1016/j.marpolbul.2023.115785
64. Bi, S., Jiang, W., & Yang, Y. (2024). Policy analysis and cooperation suggestions on ASEAN countries' marine plastic waste pollution. *Advances in Politics and Economics, 7*(3), p80. https://doi.org/10.22158/ape.v7n3p80

Chapter 4
Detection and Analysis

Introduction

Microplastics now occur across marine, freshwater, terrestrial, and atmospheric environments, which makes their reliable detection a central priority in environmental science. Accurate identification and characterization determine how researchers understand their distribution, environmental behavior, and ecological consequences. Analytical science, therefore, plays a critical role in advancing microplastic research, since conclusions on transport, exposure, and risk depend directly on measurement quality.

This chapter examines the scientific foundations of microplastic detection and analysis. It reviews the analytical tools used to isolate, identify, and quantify microplastics and nanoplastics across environmental matrices. It also evaluates methodological limitations that affect reproducibility, comparability, and data interpretation. Key issues include sampling bias, contamination risks, detection limits, and the absence of harmonized protocols. These constraints explain why reported concentrations often vary across studies and regions.

Recent technological progress continues to reshape the field. Advances in spectroscopy, thermal analysis, microscopy, and automated imaging improve polymer identification and particle characterization. Field-deployable instruments and remote sensing approaches now expand monitoring capacity beyond laboratory settings. These developments support larger spatial assessments and long-term monitoring programs.

This chapter integrates current methodological knowledge to guide the selection of robust analytical approaches. It emphasizes the importance of quality assurance, standardization, and innovation in building reliable datasets. Strong analytical foundations remain essential for advancing microplastic science, supporting environmental risk assessment, and informing evidence-based responses to plastic pollution.

R. Y. Capangpangan, A. C. Alguno, *Microplastics in Focus*,
https://doi.org/10.1007/978-3-032-28978-0_4

4.1 Tools of the Trade: Methods for Identifying Microplastics and Nanoplastics

The accurate identification of microplastics and nanoplastics in environmental matrices presents a multifaceted challenge, owing to their diversity in polymer composition, size distribution, morphology, and weathering state. Environmental samples often include sediments, biota, wastewater, and air particulates, each of which harbors particles embedded within organic or inorganic matter. The heterogeneity of these matrices requires the application of multiple analytical approaches that operate across varying scales of resolution, sensitivity, and specificity [1].

Researchers employ a sequence of processes that begin with the physical isolation of particles. Size fractionation through sieving or filtration enables the preliminary segregation of particles above and below the microplastic threshold (typically defined as 1–5 mm) and into the nanoplastic domain (<1 μm) [2]. However, operational thresholds remain arbitrary due to the lack of global consensus on size boundaries, and separation techniques often suffer from incomplete recovery, especially for particles below 10 μm. Microplastic detection typically follows a structured analytical sequence that integrates sampling, preparation, particle isolation, chemical identification, and data interpretation. Figure 4.1 illustrates this integrated workflow and highlights how different analytical stages contribute to reliable microplastic characterization. The figure illustrates the sequential analytical stages involved in microplastic research, beginning with environmental sampling, followed by sample preparation, particle sorting, polymer identification, quantitative analysis, and data integration. The workflow highlights the complementary roles of physical, spectroscopic, thermal, and data-driven techniques in improving detection reliability across environmental matrices.

After separation, visual examination under a stereomicroscope or scanning electron microscopy (SEM) provides morphological insight, though this method remains subjective and limited for smaller particles. Optical imaging often leads to misclassification, particularly for weathered plastics or particles with biofilm coatings or mineral encrustations. As such, visual inspection is insufficient as a standalone method [3].

To address this, spectroscopic methods play a central role in polymer identification. Fourier transform infrared spectroscopy (FTIR), especially in attenuated total reflectance (ATR-FTIR) and focal plane array (FPA) modes, enables non-destructive identification of particles >20 μm. Raman spectroscopy extends the resolution to 1 μm and can detect additives and pigments; however, it is susceptible to fluorescence interference, particularly in environmental samples with organic contamination [4]. Both methods rely on comparison against reference libraries, which must be standardized to reduce interlaboratory variability.

For nanoplastics, conventional spectroscopic tools reach their detection limits. Therefore, researchers increasingly integrate thermal analysis techniques. Pyrolysis-GC/MS and thermal extraction desorption-gas chromatography-mass spectrometry (TED-GC/MS) enable the chemical fingerprinting of polymers based

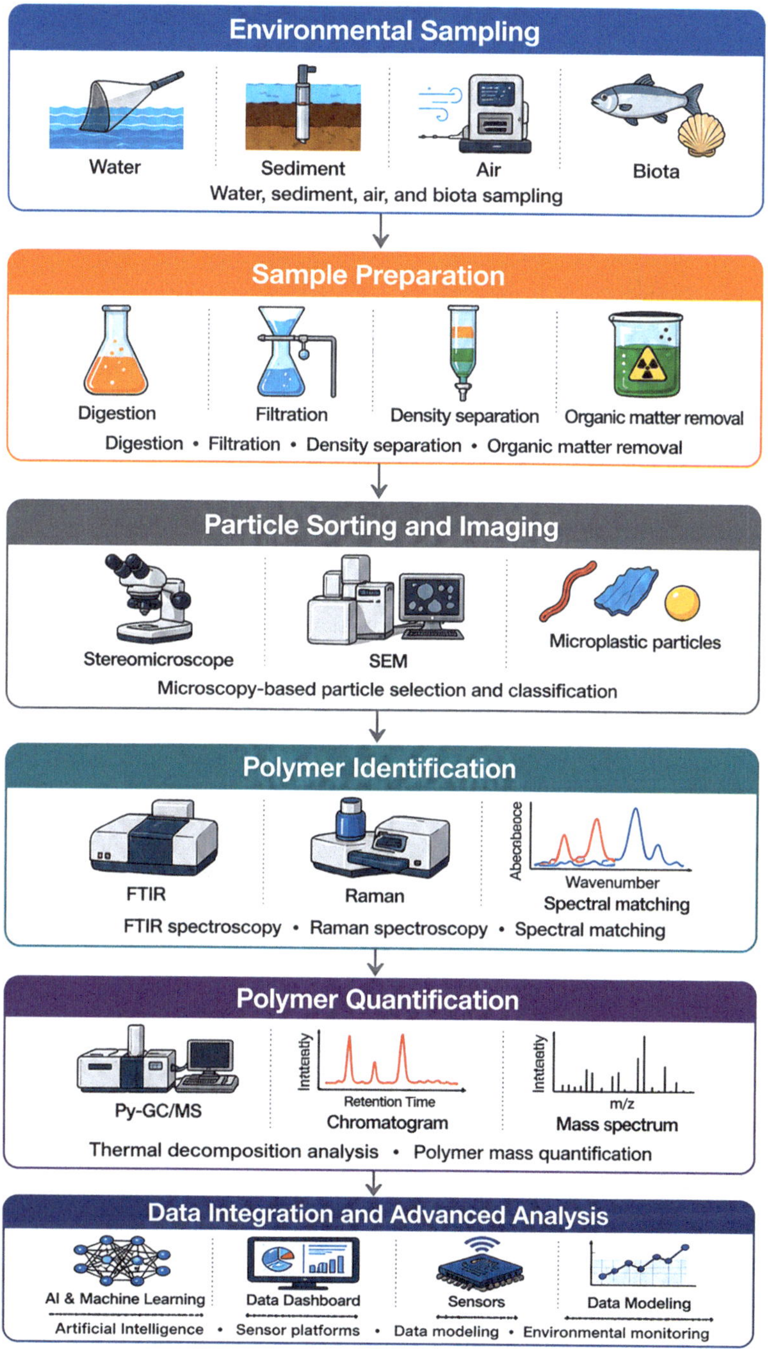

Fig. 4.1 Integrated workflow for microplastic detection and analysis. (Conceptual framework developed by the authors and illustrated using digital design tools)

on thermal decomposition patterns [5]. These techniques provide high specificity and allow quantification, yet they destroy the sample and lack morphological resolution.

Emerging techniques such as AF4 (asymmetric flow field-flow fractionation) coupled with multi-angle light scattering (MALS) or dynamic light scattering (DLS) show promise for size-resolved characterization of nanoplastics in aqueous suspensions. Nonetheless, the aggregation of nanoparticles in high-ionic-strength matrices or the presence of humic substances may alter hydrodynamic behavior and skew results [6].

Fluorescence staining using Nile Red has gained traction due to its rapid screening capabilities. When coupled with automated imaging and particle counting software, this method enhances throughput. However, fluorescence intensity varies across polymer types and depends on surface oxidation, making it semi-quantitative at best [7].

The reliability of microplastic detection results depends strongly on the analytical objective, whether focused on quantification, polymer identification, or toxicity assessment. Environmental context also plays a critical role, as different matrices require tailored extraction and preparation protocols; for example, marine sediments require different treatment approaches compared with human tissues or bottled water [8]. No single analytical technique can address all detection needs. Consequently, a multi-tiered analytical strategy that integrates physical, chemical, and thermal techniques, supported by calibration against certified reference materials, remains the current gold standard [9].

Because analytical techniques vary in detection limits, analytical outputs, and methodological constraints, comparative evaluation is essential when selecting appropriate tools. Figure 4.2 therefore compares major techniques such as FTIR, Raman spectroscopy, SEM, and Py-GC/MS in terms of detectable size range, destructive versus non-destructive analysis, polymer identification capability,

ANALYSIS METHOD	PARTICLE SIZE RANGE	DESTRUCTIVE ANALYSIS	IDENTIFICATION CAPABILITY	MORPHOLOGY DATA	QUANTIFICATION CAPABILITY
FT-IR	>20 μm	No	Yes	Limited	No
Raman	>1 μm	No	Yes	Limited	No
SEM	nm scale	No	No	Yes	No
Py-GC/MS	Bulk polymer mass	Yes	Yes	No	Yes

Yes Limited No

Fig. 4.2 Comparison of analytical techniques used in microplastic detection and analysis. (Conceptual framework developed by the authors and illustrated using digital design tools)

morphological characterization, and quantification potential. The comparison highlights how spectroscopic, microscopic, and thermal approaches complement one another to achieve comprehensive and reliable microplastic characterization.

As microplastic science advances, refining integrated analytical workflows, developing universal polymer databases, and standardizing interlaboratory validation protocols will be critical to achieving reliable detection in complex environmental settings. Together, the analytical workflow (Fig. 4.1) and the comparative overview of techniques (Fig. 4.2) demonstrate why integrated, multi-method strategies remain essential for improving detection accuracy, ensuring methodological consistency, and strengthening the scientific foundation of microplastic research.

4.1.1 Sampling and Pre-treatment

Effective detection and analysis of microplastics and nanoplastics begins with methodical and representative sampling. The variability of environmental matrices such as water, sediment, biota, and air requires customized strategies to ensure comparability and reproducibility across studies [10]. Sampling serves as the foundation for all subsequent analytical stages and must capture both spatial and temporal variability in particle distribution.

In aquatic systems, manta trawls and neuston nets are widely employed to collect floating microplastics. These devices typically use mesh sizes between 100 and 300 μm, which effectively capture particles within the microplastic size range while minimizing clogging and sampling artifacts [11]. However, these nets fail to retain smaller fragments, especially nanoplastics, which escape due to their diminutive dimensions. Therefore, researchers increasingly rely on in situ filtration systems and pump samplers capable of processing large water volumes with high-resolution filters down to 0.45 μm [12].

Sediment sampling typically requires core devices, such as grab samplers and box corers. These tools enable vertical stratification and depth-specific sampling, which are essential for understanding the dynamics of historical deposition and resuspension [13]. For atmospheric microplastics, high-volume air samplers and passive deposition collectors are used to capture airborne fibers and fragments, including both inhalable and settleable fractions, which vary with urban density and meteorological conditions [14]. Because environmental matrices require compartment-specific strategies, researchers rely on sampling devices tailored to whether the target particles are in water, sediments, or air.

Following sampling, environmental materials must undergo pre-treatment to isolate microplastics from organic and mineral matrices. Wet peroxide oxidation (WPO), employing hydrogen peroxide catalyzed by iron, is widely used to digest natural organics while preserving polymer integrity [15]. Enzymatic digestion offers a milder alternative, particularly suited for biota samples, where its selectivity improves recovery of internalized particles [16]. These procedures enhance polymer visibility and prevent interference during spectroscopic or microscopic detection.

4.1.2 Virtual Sorting and Microscopy

Accurate detection of microplastics in environmental matrices often begins with visual inspection under stereomicroscopes. These instruments enable rapid screening of particles larger than 300 μm based on observable traits, such as shape, color, and surface texture [10]. While stereomicroscopy provides a valuable preliminary assessment, its reliance on human interpretation introduces subjectivity, increasing the likelihood of misclassification, especially when differentiating microplastics from natural organic or mineral debris [7].

To address the limitations of visual techniques, researchers increasingly turn to scanning electron microscopy (SEM) for high-resolution surface imaging. SEM provides detailed morphological data at the micro- and nanoscale, revealing features such as polymer fissures, weathering marks, and surface roughness that are often imperceptible through optical methods [17]. The ability of SEM to distinguish between synthetic and non-synthetic particulates enhances analytical confidence, especially when used in conjunction with elemental analysis tools, such as energy-dispersive X-ray spectroscopy (EDS) [18]. This integration enables both morphological and compositional confirmation, which is essential when analyzing heterogeneous environmental samples. Together, stereomicroscopy and SEM serve as foundational components in the analytical toolbox for microplastic research, providing complementary perspectives for robust particle identification and classification in complex matrices.

4.1.3 Spectroscopic Techniques

Fourier transform infrared spectroscopy and Raman spectroscopy form the analytical backbone of polymer identification in microplastic research. These vibrational spectroscopic techniques enable direct chemical characterization by detecting molecular bond vibrations specific to particular polymer structures. Micro-Fourier transform infrared spectroscopy, particularly when combined with focal-plane-array detectors, enables automated mapping and identification of particles as small as 20 μm across large sample areas with high throughput and reproducibility [19]. This capability makes FTIR well-suited for routine screening of environmental samples, such as sediments, wastewater residues, and biota extracts.

Raman spectroscopy extends analytical resolution to the submicrometer scale and supports the detection of nanoplastics below 1 μm. Its higher spatial resolution enables detailed analysis of small fragments, fibers, and weathered particles embedded within complex matrices [17]. Raman analysis also detects pigments and additives, which aid in tracing plastic sources and degradation pathways. However, fluorescence interference from organic matter and surface biofilms can complicate spectral interpretation.

Both techniques rely on comparison with validated spectral libraries to assign polymer identity. The accuracy of this matching process depends on library completeness, spectral quality, and standardized acquisition protocols [20]. When combined, FTIR and Raman spectroscopy offer complementary strengths, enabling robust polymer identification across a wide size range and supporting reliable microplastic classification in environmental studies.

4.1.4 Thermal Analysis

Pyrolysis-gas chromatography/mass spectrometry (Py-GC/MS) is among the most robust techniques for polymer-specific quantification of microplastics, especially in complex matrices, such as sediments, sewage sludge, and biota [21]. This method operates by thermally decomposing polymeric materials into characteristic pyrolyzates under oxygen-free conditions, followed by separation via gas chromatography and detection through mass spectrometry [22]. These pyrolyzates act as molecular fingerprints, enabling accurate polymer identification and quantification even in samples heavily contaminated with organic matter or mineral particles [23].

One of Py-GC/MS's notable strengths is its ability to process heterogeneous environmental samples without the need for extensive pre-separation or filtration. It can directly quantify total polymer mass, which is essential for risk assessment and flux estimation in aquatic systems [24]. When combined with isotopically labeled internal standards, the method achieves high precision and reproducibility [5]. However, a key limitation is its inability to provide information on particle size, shape, or count. Since the analysis focuses on bulk polymer mass, it lacks spatial resolution and does not differentiate between microplastics and nanoplastics or their morphological states [12].

Despite this constraint, Py-GC/MS has proven indispensable for studies requiring accurate quantification of polymer loads in environmental samples with high background noise. Its application in sewage sludge and soil has revealed significant accumulation of synthetic polymers in terrestrial sinks, underscoring its role in fate and transport studies [25]. Researchers have also adapted the technique to target specific polymer markers, enhancing detection limits and reducing false positives in mixed-material matrices [26]. These refinements are essential as regulatory frameworks increasingly require standardized, validated methods for environmental monitoring and product safety compliance.

Py-GC/MS complements spectroscopic techniques such as FTIR and Raman by bridging the quantification gap, particularly in samples unsuitable for optical analysis. As environmental microplastic research advances toward holistic exposure assessments, integrating Py-GC/MS with imaging and spectroscopic workflows is becoming a best practice in multidisciplinary studies [27].

4.1.5 *Emerging Techniques*

Nano-FTIR and atomic force microscopy-infrared spectroscopy (AFM-IR) offer spatial resolution of 20 nm, ideal for nanoscale plastics [28]. Fluorescence tagging with Nile Red enables rapid, semi-quantitative screening but may yield false positives due to nonspecific binding [7].

Multimodal approaches combining spectroscopy and thermal analysis improve confidence in detection. For instance, coupling FTIR with Py-GC/MS confirms both particle morphology and polymer identity, enhancing reliability across matrices [29].

4.2 Challenges in Microplastic Research: Accuracy, Scale, and Standardization

Analytical advancements have led to deeper insights into microplastic pollution, yet several challenges persist. Three core issues affect the reliability and comparability of data: accuracy, scale, and standardization. These interconnected challenges underpin current limitations in microplastic detection and collectively determine the reproducibility, comparability, and regulatory usability of reported datasets.

Figure 4.3 presents a conceptual framework that synthesizes these three analytical dimensions—analytical accuracy, measurement scale, and methodological

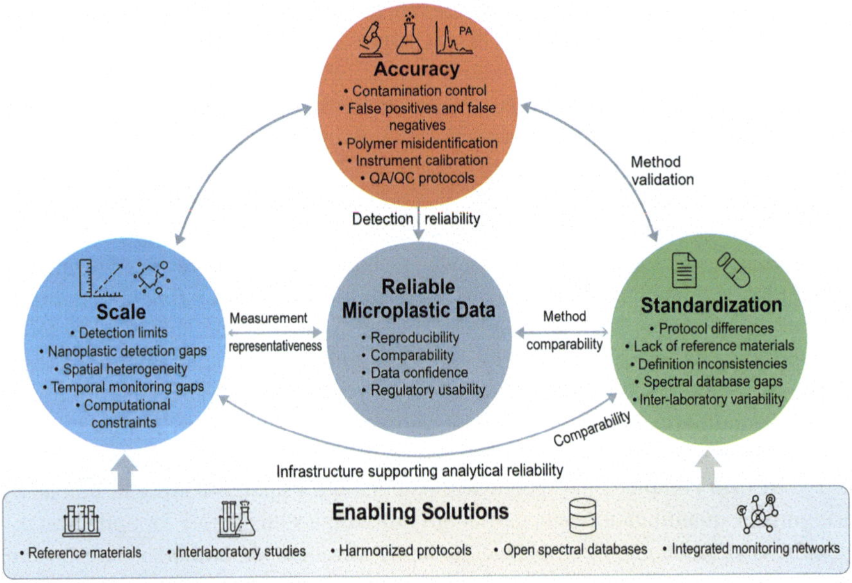

Fig. 4.3 Core analytical challenges affecting the reliability of microplastic research. (Conceptual framework developed by the authors and illustrated using digital design tools)

standardization—and illustrates how their interactions shape the overall reliability of microplastic research. Accuracy involves contamination control and correct polymer identification; scale reflects particle-size detection limits and environmental heterogeneity; and standardization addresses protocol harmonization and the availability of reference materials. Together, these factors determine the reproducibility, comparability, and scientific reliability of microplastic datasets.

4.2.1 Accuracy

Accuracy in microplastic detection requires minimizing false positives and false negatives. Visual sorting often misidentifies natural fibers or organic debris as plastics [30]. Spectroscopic analysis depends heavily on the quality of spectral libraries and instrument calibration [12].

Nanoplastics, due to their size and altered physicochemical properties, elude most traditional detection methods. They often exhibit signal interference, background fluorescence, and aggregation during sample processing. Fluorescent tagging techniques lack specificity when other organic compounds bind dyes like Nile Red [7].

Thermal degradation methods, while highly sensitive, cannot differentiate between polymers of identical chemical composition but varying structure or origin. Moreover, environmental contamination during sampling and laboratory processing further compromises data fidelity [15].

Quality assurance protocols involving blanks, spike-and-recovery tests, and procedural replicates enhance data accuracy. Standard reference materials for microplastics remain limited, hampering interlaboratory comparisons [31].

4.2.2 Scale

Scale refers to both the size of particles analyzed and the spatial-temporal breadth of monitoring. Traditional sampling tools fail to retain nanoplastics and particles smaller than 10 μm. FTIR and Raman mapping of large datasets demand significant time and computational resources [32].

Spatial heterogeneity of microplastics complicates representative sampling across regions and ecosystems. Variability in hydrodynamics, land use, and population density influences microplastic abundance and type [11].

Temporal monitoring remains inconsistent, with most studies providing snapshots rather than long-term trends. High-resolution monitoring networks that integrate multiple compartments (air, soil, biota, and water) remain rare, hindering risk assessment [33].

4.2.3 Standardization

Lack of methodological harmonization undermines comparability among studies. Differences in mesh sizes, sample volumes, digestion methods, and particle classification criteria produce divergent results [10].

The absence of unified definitions for nanoplastics also impedes regulatory assessments. While microplastics are generally defined as particles <5 mm, the lower threshold for nanoplastics is ambiguously reported as 1 μm–1 nm across studies [6].

Efforts such as the European Commission's JRC Technical Reports and the United Nations Group of Experts on the Scientific Aspects of Marine Environmental Protection (GESAMP) promote standard protocols. However, global implementation remains uneven [34].

Research networks and interlaboratory comparison exercises play vital roles in validating and refining methods. Continued development of reference materials, certified standards, and open-access spectral databases is critical for methodological convergence [35].

4.3 Innovations in Monitoring: New Frontiers in Field-Based and Remote Sensing

Recent years have seen a paradigm shift in monitoring microplastics, driven by innovations in real-time, field-deployable technologies. These emerging tools aim to reduce processing time, expand monitoring coverage, and enable high-frequency data collection.

4.3.1 Portable Spectroscopic Devices

Handheld Raman and FTIR devices allow in situ polymer identification. They provide rapid feedback on particle composition, facilitating on-site decision-making during field campaigns [20]. Although resolution remains lower than that of lab-based systems, technological refinements continue to improve their reliability.

4.3.2 Remote Sensing and Hyperspectral Imaging

Hyperspectral sensors capture reflectance data across hundreds of wavelengths, enabling differentiation of plastic types on beaches and in surface waters. Drone-mounted platforms and satellites equipped with hyperspectral cameras allow large-area surveillance of visible macro- and mesoplastics [36].

4.3.3 Machine Learning Integration

Artificial intelligence (AI) applications in image recognition and spectral classification enhance the speed and accuracy of particle identification. Deep learning algorithms trained on annotated datasets automate detection, reducing human error [37].

4.3.4 Microfluidic and Sensor-Based Platforms

Microfluidic chips embedded with optical sensors separate and detect microplastics from small volumes of water. These platforms integrate sample preparation, separation, and detection in a compact system suitable for environmental monitoring [38].

4.3.5 eDNA and Biomarker Approaches

Environmental DNA methods trace the presence of microplastics by identifying associated microbial communities or pollutant biomarker profiles. This indirect detection offers ecological context, particularly for nanoplastics and ingested particles [39, 40].

4.3.6 Autonomous Samplers and Smart Buoys

Autonomous underwater vehicles and smart buoys equipped with sampling modules collect water and sediment samples in real-time. They support continuous monitoring across remote and dynamic systems, such as estuaries and coastal zones [41].

Continued innovation in these areas will enable integration of multi-source data into spatial models and risk assessment frameworks, ultimately strengthening global efforts to monitor and manage microplastic pollution.

4.4 Chapter Synthesis

This chapter demonstrates that analytical capability defines the strength of microplastic science. Progress in understanding environmental distribution and ecological risk depends on the accuracy, sensitivity, and consistency of detection methods. The discussion shows that no single analytical approach addresses all research needs. Instead, researchers must combine complementary techniques to achieve reliable identification and quantification.

The evaluation of current methodologies shows clear trade-offs between resolution, throughput, cost, and accessibility. Spectroscopic techniques such as FTIR and Raman spectroscopy provide strong evidence of polymer identity but require careful sample preparation. Thermal methods allow mass-based quantification but sacrifice particle-level detail. Microscopy-based approaches support morphological characterization but require confirmation through chemical analysis. These methodological realities require careful study design to ensure credible results.

This chapter also highlights the need for standardization. Differences in sampling strategies, processing protocols, and reporting units continue to limit cross-study comparisons. Harmonized protocols and interlaboratory validation remain necessary to improve global datasets. Emerging monitoring technologies offer strong potential to address these limitations through automation, artificial intelligence-assisted analysis, and expanded field deployment.

Taken together, this chapter establishes that detection science underpins microplastic research. Without reliable analytical methods, environmental assessments remain uncertain, and risk evaluations remain incomplete. Analytical innovation, therefore, remains a prerequisite for strengthening scientific confidence, guiding policy responses, and advancing mitigation strategies.

Scientific detection explains where microplastics occur and how they can be measured. The next question concerns what happens after exposure occurs. Once microplastics enter ecosystems, they interact with organisms, food webs, and ecological processes. These interactions determine biological responses, ecosystem stability, and potential threats to biodiversity.

The next chapter examines the ecological consequences of microplastic pollution. It evaluates their effects across marine and freshwater ecosystems, their accumulation in soils and agricultural environments, and their influence on biodiversity and trophic interactions. Building on advances in analytical understanding, the discussion moves from detection and measurement toward biological responses and ecosystem-level effects.

References

1. Prata, J. C., da Costa, J. P., Lopes, I., Duarte, A. C., & Rocha-Santos, T. (2019). Methods for sampling and detection of microplastics in water and sediment: A critical review. *TrAC Trends in Analytical Chemistry, 110*, 150–159. https://doi.org/10.1016/j.trac.2018.10.029
2. Ivleva, N. P., Wiesheu, A. C., & Niessner, R. (2017). Microplastic in aquatic ecosystems. *Angewandte Chemie International Edition, 56*(7), 1720–1739. https://doi.org/10.1002/anie.201606957
3. Masura, J., Baker, J. E., Foster, G. D., Arthur, C., & Herring, C. (2015). Laboratory methods for the analysis of microplastics in the marine environment. *NOAA Technical Memorandum NOS-OR&R-48*. https://marinedebris.noaa.gov/sites/default/files/publications-files/noaa_microplastics_methods_techmemo2015.pdf
4. Anger, P. M., von der Esch, E., Baumann, T., Elsner, M., Niessner, R., & Ivleva, N. P. (2018). Raman microspectroscopy as a tool for microplastic particle analysis. *TrAC Trends in Analytical Chemistry, 109*, 214–226. https://doi.org/10.1016/j.trac.2018.10.010

5. Dümichen, E., Eisentraut, P., Bannick, C. G., Barthel, A. K., Senz, R., & Braun, U. (2017). Fast identification of microplastics in complex environmental samples by a thermal degradation method. *Chemosphere, 174*, 572–584. https://doi.org/10.1016/j.chemosphere.2017.02.010
6. Gigault, J., ter Halle, A., Baudrimont, M., Pascal, P.-Y., Gauffre, F., Phi, T.-L., El Hadri, H., Grassl, B., & Reynaud, S. (2018). Current opinion: What is a nanoplastic? *Environmental Pollution, 235*, 1030–1034. https://doi.org/10.1016/j.envpol.2018.01.024
7. Erni-Cassola, G., Zadjelovic, V., Gibson, M. I., & Christie-Oleza, J. A. (2017). Distribution of plastic polymer types in the marine environment; a meta-analysis. *Journal of Hazardous Materials, 369*, 691–698. https://doi.org/10.1016/j.jhazmat.2019.02.067
8. Shim, W. J., Hong, S. H., & Eo, S. (2018). Marine microplastics: Abundance, distribution, and composition. In E. Y. Zeng (Ed.), *Microplastic contamination in aquatic environments: An emerging matter of environmental urgency* (pp. 1–26). Elsevier. https://doi.org/10.1016/B978-0-12-813747-5.00001-1
9. Wesch, C., Elert, A. M., Wörner, M., Braun, U., Klein, R., & Paulus, M. (2016). Assuring quality in microplastic monitoring: About the value of clean-air devices as essentials for verified data. *Scientific Reports, 7*, 5424. https://doi.org/10.1038/s41598-017-05838-4
10. Hidalgo-Ruz, V., Gutow, L., Thompson, R. C., & Thiel, M. (2012). Microplastics in the marine environment: A review of the methods used for identification and quantification. *Environmental Science & Technology, 46*(6), 3060–3075. https://doi.org/10.1021/es2031505
11. Lindeque, P. K., Cole, M., Coppock, R. L., Lewis, C. N., Miller, R. Z., Watts, A. J. R., Wilson-McNeal, A., Wright, S. L., & Galloway, T. S. (2020). Are we underestimating microplastic abundance in the marine environment? A comparison of microplastic capture with nets of different mesh-size. *Environmental Pollution., 265*(Part A), 114721. https://doi.org/10.1016/j.envpol.2020.114721
12. Hale, R. C., Seeley, M. E., King, A. E., & Yu, L. H. (2021). Analytical chemistry of plastic debris: Sampling, methods, and instrumentation. In M. S. Bank (Ed.), *Microplastic in the environment: Pattern and process* (pp. 17–67). Springer. https://doi.org/10.1007/978-3-030-78627-4_2
13. Claessens, M., Van Cauwenberghe, L., Vandegehuchte, M. B., & Janssen, C. R. (2011). New techniques for the detection of microplastics in sediments and field collected organisms. *Marine Pollution Bulletin, 70*(1–2), 227–233. https://doi.org/10.1016/j.marpolbul.2013.03.009
14. Allen, S., Allen, D., Phoenix, V. R., Le Roux, G., Durántez Jiménez, P., Simonneau, A., Binet, S., & Galop, D. (2019). Atmospheric transport and deposition of microplastics in a remote mountain catchment. *Nature Geoscience, 12*(5), 339–344. https://doi.org/10.1038/s41561-019-0335-5
15. Tagg, A. S., Sapp, M., Harrison, J. P., & Ojeda, J. J. (2017). Identification and quantification of microplastics in wastewater using focal plane array-based reflectance micro-FT-IR imaging. *Analytical Chemistry, 87*(12), 6032–6040. https://doi.org/10.1021/acs.analchem.5b00495
16. Cole, M., Lindeque, P., Fileman, E., Halsband, C., & Galloway, T. S. (2014). The impact of polystyrene microplastics on feeding, function and fecundity in the marine copepod *Calanus helgolandicus*. *Environmental Science & Technology, 49*(2), 1130–1137. https://doi.org/10.1021/es504525u
17. Xu, J. L., Thomas, K. V., Luo, Z., & Gowen, A. A. (2019). FTIR and Raman imaging for microplastics analysis: State of the art, challenges and prospects. *TrAC Trends in Analytical Chemistry, 119*, 115629. https://doi.org/10.1016/j.trac.2019.115629
18. Käppler, A., Windrich, F., Löder, M. G. J., Malanin, M., Fischer, D., Labrenz, M., Eichhorn, K. J., & Voit, B. (2015). Identification of microplastics by FTIR and Raman microscopy: A novel silicon filter substrate opens the important spectral range below 1300 cm−1 for FTIR transmission measurements. *Analytical and Bioanalytical Chemistry, 407*(22), 6791–6801. https://doi.org/10.1007/s00216-015-8850-8
19. Li, J., Liu, H., & Chen, J. P. (2018). Microplastics in freshwater systems: A review on occurrence, environmental effects, and methods for microplastics detection. *Water Research, 137*, 362–374. https://doi.org/10.1016/j.watres.2017.12.056
20. Löder, M. G. J., Kuczera, M., Mintenig, S., Lorenz, C., & Gerdts, G. (2015). Focal plane array detector based micro Fourier transform infrared imaging for the analysis

of microplastics in environmental samples. *Environmental Chemistry, 12*(5), 563–581. https://doi.org/10.1071/EN14205
21. Peel, R. H., Lloyd, C., Roberts, S., Naafs, B. D. A., & Bull, I. D. (2024). Quantification of microplastic targets in environmental matrices using pyrolysis-gas chromatography-mass spectrometry. *Environmental Science Advances, 4*(1), 159. https://doi.org/10.1039/d4va00269e
22. Picó, Y., & Barceló, D. (2020). Pyrolysis gas chromatography-mass spectrometry in environmental analysis: Focus on organic matter and microplastics. *TrAC Trends in Analytical Chemistry, 130*, 115964. https://doi.org/10.1016/j.trac.2020.115964
23. Santos, L. H. M. L. M., Insa, S., Arxé, M., Buttiglieri, G., Rodríguez-Mozaz, S., & Barcelo, D. (2023). Analysis of microplastics in the environment: Identification and quantification of trace levels of common types of plastic polymers using pyrolysis-GC/MS. *MethodsX, 10*, 102143. https://doi.org/10.1016/j.mex.2023.102143
24. Dalmau-Soler, J., Boleda, M. R., & Lacorte, S. (2024). Routine method for the analysis of microplastics in natural and drinking water by pyrolysis coupled to gas chromatography-mass spectrometry. *Journal of Chromatography A, 1730*, 465153. https://doi.org/10.1016/j.chroma.2024.465153
25. Okoffo, E. D., Rauert, C., & Thomas, K. V. (2023). Mass quantification of microplastic at wastewater treatment plants by pyrolysis-gas chromatography–mass spectrometry. *Science of the Total Environment, 856*(Part 2), 159251. https://doi.org/10.1016/j.scitotenv.2022.159251
26. Singh, J. (2023). Advanced analytical techniques for microplastics in the environment: A review. *Bulletin of the National Research Centre/Bulletin of the National Research Center, 47*(1), 174. https://doi.org/10.1186/s42269-023-01148-0
27. Huang, Z., Hu, B., & Wang, H. (2023). Analytical methods for microplastics in the environment: A review. *Environmental Chemistry Letters, 21*, 383–401. https://doi.org/10.1007/s10311-022-01525-7
28. Xie, J., Gowen, A., Xu, W., & Xu, J. (2024). Analysing micro- and nanoplastics with cutting-edge infrared spectroscopy techniques: A critical review. *Analytical Methods, 16*(15), 2177. https://doi.org/10.1039/d3ay01808c
29. Yu, J., Wang, P., Ni, F., Cizdziel, J. V., Wu, D., Zhao, Q., & Zhou, Y. (2019). Characterization of microplastics in environment by thermal gravimetric analysis coupled with Fourier transform infrared spectroscopy. *Marine Pollution Bulletin, 145*, 153. https://doi.org/10.1016/j.marpolbul.2019.05.037
30. Löder, M. G. J., & Gerdts, G. (2015). Methodology used for the detection and identification of microplastics—A critical appraisal. In *Marine anthropogenic litter* (p. 201). Springer. https://doi.org/10.1007/978-3-319-16510-3_8
31. Brander, S. M., Renick, V. C., Foley, M. M., Steele, C., Woo, M., Lusher, A., Carr, S., Helm, P. A., Box, C., Cherniak, S. L., Andrews, R. C., & Rochman, C. M. (2020). Sampling and quality assurance and quality control: A guide for scientists investigating the occurrence of microplastics across matrices. *Applied Spectroscopy, 74*(9), 1099. https://doi.org/10.1177/0003702820945713
32. Liang, J., Ling, J., Xu, L., & Zhu, X. (2025). A dataset of Raman and infrared spectra as an extension to the ChEMBL. *Scientific Data, 12*, 939. https://doi.org/10.1038/s41597-025-03939-2
33. Savoca, M. S., Abreo, N. A., Arias, A. H., Baes, L., Baini, M., Bergami, E., Brander, S., Canals, M., Choy, C. A., Corsi, I., De Witte, B., Domit, C., Dudas, S., Duncan, E. M., Fernández, C. E., Fossi, M. C., Garcés-Ordóñez, O., Godley, B. J., González-Paredes, D., ... Santos, R. G. (2025). Monitoring plastic pollution using bioindicators: A global review and recommendations for marine environments. *Environmental Science: Advances, 4*, 10–32. https://doi.org/10.1039/D4VA00174E
34. Watson-Wright, W. M., Wells, P. G., Duce, R. A., Gilardi, K. V., Girvan, S. T. G., Huber, E. L., Kershaw, P. J., Linders, J. B. H. J., Luit, R., Viviani, C. M. G., & Housden, D. V. (2024). The UN Joint Group of Experts on the Scientific Aspects of Marine Environmental Protection (GESAMP)—An ocean science–policy interface standing the test of time. *Marine Pollution Bulletin, 199*, 115917. https://doi.org/10.1016/j.marpolbul.2023.115917

35. Rognerud, I., Hurley, R., Lusher, A., Bråte, I. L. N., & Steindal, E. H. (2022). Addressing microplastics in a global agreement on plastic pollution. In *TemaNord*. https://doi.org/10.6027/temanord2022-566
36. Chabot, D., & Marteinson, S. C. (2024). Aerial remote sensing of aquatic microplastic pollution: The state of the science and how to move it forward. *Microplastics, 3*(4), 685. https://doi.org/10.3390/microplastics3040042
37. Arju, M. Z. B. Z., Hridi, N. A., Dewan, L., Suhaila, S., Amin, M. N., Rashid, T. U., Azad, A. K., Rahman, S., Hossain, M., & Habib, A. (2025). Deep-learning enabled rapid and low-cost detection of microplastics in consumer products following on-site extraction and image processing. *RSC Advances, 15*(14), 10473. https://doi.org/10.1039/d4ra07991d
38. Paruli, E. I., Sainte-Suzanne, A. D. L., Debeaumont, M., Thomas, L., Courson, R., Challier, L., Rakwe, M. E., & Prado, É. (2025). In-flow single particle detection of sub-100 micron microplastics. *RSC Advances, 15*(40), 33139. https://doi.org/10.1039/d5ra04700e
39. Wani, A. K., Akhtar, N., Naqash, N., Rahayu, F., Djajadi, D., Chopra, C., Singh, R., Mulla, S. I., Sher, F., & Américo-Pinheiro, J. H. P. (2023). Discovering untapped microbial communities through metagenomics for microplastic remediation: Recent advances, challenges, and way forward. *Environmental Science and Pollution Research, 30*(34), 81450. https://doi.org/10.1007/s11356-023-25192-5
40. Wu, N. (2022). Tracing microplastic footprints through the plastisphere. *Nature Reviews Earth & Environment, 3*(8), 498. https://doi.org/10.1038/s43017-022-00321-9
41. Shukla, A., Ross, R., Bhattacharya, B., & Stumpf, A. (2025). Autonomous water sampling and quality monitoring in remote locations: A novel approach using a remote-controlled boat. *HardwareX, 22*, e00634. https://doi.org/10.1016/j.ohx.2025.e00634

Part II
The Impact of Microplastics

Chapter 5
Ecological Impacts

Introduction

Microplastics and nanoplastics are now recognized as pervasive contaminants in terrestrial, freshwater, and marine environments, with implications that extend far beyond localized pollution. These synthetic particles, typically less than 5 mm in size, exert complex ecological pressures on living systems across trophic levels, from plankton to apex predators [1, 2]. Their ubiquity in sediment, surface water, and even the atmosphere signals a shift in environmental paradigms, in which plastics function not only as physical pollutants but also as chemical vectors that can disrupt ecological integrity [3].

This chapter synthesizes current scientific insights into how microplastics and nanoplastics affect aquatic and terrestrial ecosystems, soil health, biodiversity, and food webs, with particular emphasis on pathways linked to food security in the Global South. Emphasis is placed on sublethal and chronic effects that alter organismal physiology, reproductive output, feeding behavior, and interspecies dynamics [4]. In freshwater systems, impacts manifest differently than in marine zones due to variable hydrodynamics and species composition [5]. Soil and agricultural contamination, a more recent frontier of concern, shows that plastics can hinder plant growth, disrupt microbiomes, and diminish soil fertility [6].

The subsequent sections delve into specific environments and biological responses, guided by quantitative evidence and ecotoxicological frameworks. Through this approach, this chapter aims to inform policies, monitoring systems, and long-term strategies that mitigate the ecological risks associated with plastic pollution.

R. Y. Capangpangan, A. C. Alguno, *Microplastics in Focus*,
https://doi.org/10.1007/978-3-032-28978-0_5

5.1 Effects on Marine and Freshwater Ecosystems

The infiltration of microplastics and nanoplastics into marine and freshwater ecosystems has emerged as a critical environmental concern, reshaping how scientists evaluate ecological health and biogeochemical interactions. These particles, originating from synthetic fibers, degraded consumer plastics, and industrial discharge, pervade all aquatic compartments—from surface waters to sediments and benthic zones. Their ecological effects are both direct and systemic, affecting physiological processes, trophic structures, and ecosystem resilience.

5.1.1 *Prevalence and Transport Mechanisms*

Microplastics enter aquatic systems through multiple pathways, including wastewater effluents, household and industrial discharges, agricultural runoff, shipping activities, and atmospheric deposition. Once introduced, their environmental fate is governed by hydrodynamic conditions, polymer density, particle size, and biological interactions, such as biofilm formation [7]. In marine environments, buoyant polymers such as polyethylene and polypropylene tend to accumulate in surface waters, whereas denser particles and biofouled plastics are more likely to settle in benthic sediments [8]. In freshwater systems, particularly rivers and lakes, microplastics interact with suspended sediments and organic matter, facilitating both horizontal transport along flow paths and vertical redistribution through sedimentation and resuspension processes [9, 10]. Figure 5.1 synthesizes these interconnected processes by illustrating the major entry pathways of primary and secondary microplastics, key environmental processes such as sedimentation, resuspension, ingestion, and egestion by aquatic organisms, bioaccumulation, contaminant leaching, and interactions with organic matter, as well as their ecological consequences, including trophic transfer to wildlife and humans through aquatic food consumption and drinking water. Together, these pathways highlight the interconnected nature of aquatic pollution, ecosystem exposure, and potential human health risks.

Studies in major estuarine systems, such as the Yangtze River and the Mekong Basin, reveal alarming concentrations of microplastics, often exceeding hundreds of thousands of particles per cubic meter, with fibers and fragments as the most dominant morphotypes [12, 13]. These dynamics significantly affect sediment quality, aquatic food webs, and biogeochemical cycling, particularly through the accumulation of microplastics and associated contaminants in benthic environments, followed by their transfer through aquatic organisms to higher trophic levels.

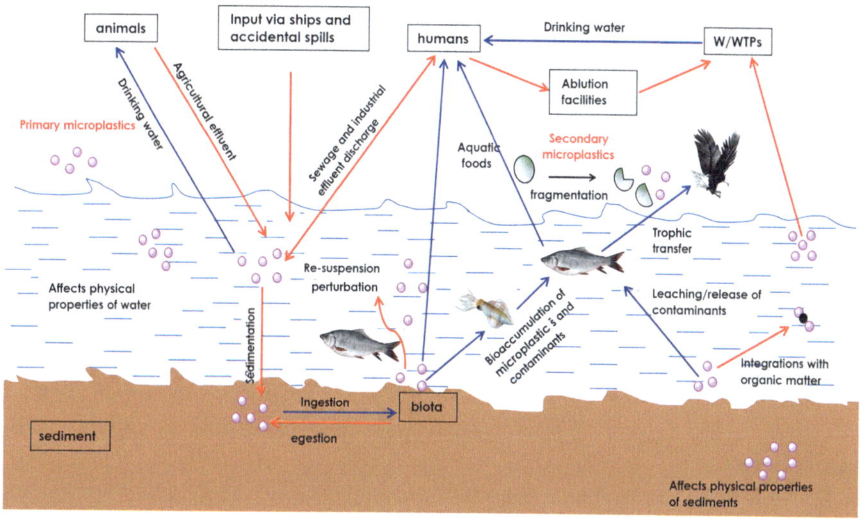

Fig. 5.1 Sources, transport pathways, environmental interactions, and trophic transfer of microplastics in aquatic ecosystems. (Adapted from [11]. This figure reproduced under the Creative Commons Attribution 4.0 International License (CC BY 4.0) (https://creativecommons.org/licenses/by/4.0/))

5.1.2 *Bioavailability and Ingestion by Aquatic Organisms*

Aquatic organisms at all trophic levels ingest microplastics, mistaking them for prey or consuming them passively during filter feeding. Zooplankton, which serve as the base of the aquatic food web, show impaired feeding rates, reduced reproductive success, and altered lipid metabolism when exposed to microplastic suspensions [14, 15]. Larval and juvenile stages of fish and bivalves are particularly vulnerable due to incomplete digestive systems and limited mobility [16].

In benthic zones, deposit feeders such as polychaetes and holothurians accumulate microplastics through sediment ingestion. Exposure experiments in *Lumbriculus variegatus* and *Arenicola marina* have shown intestinal inflammation, reduced burrowing activity, and energy depletion [17]. These effects cascade upward, influencing predator-prey interactions and overall population fitness.

5.1.3 *Physiological and Biochemical Disruptions*

Microplastics exert adverse physiological impacts by physically abrading the digestive lining and releasing adsorbed contaminants. Polymers such as polystyrene and polyvinyl chloride leach endocrine-disrupting chemicals like bisphenol A and phthalates [18, 19]. Adsorbed persistent organic pollutants (POPs), including PCBs and PAHs, amplify toxicity, causing oxidative stress, hepatocellular damage, and immunomodulation in fish and crustaceans [20, 21].

Mussels exposed to environmentally relevant concentrations of microplastics show lysosomal destabilization, altered enzymatic activities, and changes in byssus production [22]. These biochemical markers provide early warning signals of sublethal stress and ecosystem instability.

5.1.4 Trophic Transfer and Bioaccumulation

Microplastics do not degrade in the digestive tract, allowing their transfer across trophic levels. Laboratory and mesocosm studies have documented the movement of microplastics from plankton to fish and from prey fish to top predators, such as seabirds and marine mammals [23, 24]. Although biomagnification remains under investigation, the repeated ingestion across food webs raises concerns about cumulative effects.

In estuarine systems, the co-occurrence of microplastics and algal blooms exacerbates hypoxia, compromising benthic fauna and reducing nursery functions for commercially important fish [25, 26]. Trophic dilution of energy reserves due to plastic ingestion weakens food web stability, particularly in oligotrophic environments where prey availability is already limited.

5.1.5 Sediment Interactions and Benthic Ecology

Microplastics accumulate in sediments, affecting porosity, nutrient exchange, and microbial assemblages. Their presence alters denitrification rates and carbon cycling, with downstream effects on water quality [27]. Sediment-dwelling invertebrates suffer mechanical damage and exhibit altered behavior, disrupting ecosystem engineering roles vital for nutrient resuspension and oxygenation [28].

Mangrove and seagrass systems are not immune. The entrapment of microplastics in root zones reduces sediment cohesion and the potential for carbon sequestration. These impacts threaten blue carbon ecosystems, which are critical buffers against climate change and coastal erosion.

5.1.6 Freshwater Case Studies and Gaps

Freshwater environments remain underexplored, despite emerging evidence that microplastics are ubiquitous in urban lakes, agricultural canals, and reservoirs. The Rhine, Danube, and Great Lakes have shown significant contamination levels, with surface waters containing up to 400,000 particles/km^2 [29]. Microplastics alter macroinvertebrate assemblages, reduce diversity indices, and shift energy flow dynamics [30].

Field surveys and biomonitoring programs in developing countries remain scarce. In regions of the Global South, limited wastewater infrastructure and mismanaged plastic waste heighten exposure risk, yet baseline data are lacking. Future assessments must integrate spatiotemporal sampling, community-based monitoring, and transdisciplinary tools to fill these knowledge gaps.

5.2 Microplastics in Soil and Agricultural Systems

Microplastics (MPs) have become pervasive across terrestrial systems, including agricultural soils, where their presence increasingly threatens soil health, crop productivity, and food safety. While research on marine pollution has progressed rapidly, terrestrial-microplastic pollution remains less explored, despite soils being potential long-term sinks for these persistent pollutants [31]. Agricultural systems represent a unique case where deliberate and incidental inputs of plastics, such as plastic mulch films, compost, irrigation water, biosolids, and atmospheric fallout, serve as dominant vectors of microplastic entry [32].

5.2.1 Pathways of Microplastics into Agroecosystems

Multiple studies have documented that plastic mulch degradation is a key contributor to microplastic accumulation in soils. Prolonged exposure to UV radiation, tillage, and microbial activity gradually fragments these materials into micro-sized residues [33]. Furthermore, land application of sewage sludge, often rich in synthetic fibers and fragments, introduces high loads of MPs into farmlands [34]. Organic fertilizers, especially compost derived from municipal solid waste, also act as an unregulated pathway for MPs [35].

Atmospheric deposition has also emerged as a significant pathway for microplastic contamination in agroecosystems. Dris et al. [36] documented the ongoing settling of airborne microfibers in both urban and rural environments, demonstrating that even geographically remote agricultural areas are not immune to plastic fallout. In parallel, irrigation with treated or untreated wastewater represents another important entry route, introducing both microplastics and nanoplastics that frequently bypass conventional filtration systems [37]. As illustrated in Fig. 5.2, agricultural soils receive microplastics from multiple sources, including sewage sludge application, compost amendments, degradation of plastic mulching materials, irrigation inputs, and atmospheric deposition.

Following their introduction, microplastics interact with soil physicochemical processes, altering soil aggregation, facilitating pollutant adsorption and transport, and inducing shifts in microbial community structure. These changes may disrupt nutrient cycling and the ecological balance of the soil. The figure further highlights how these particles interact with soil biota, including earthworms and

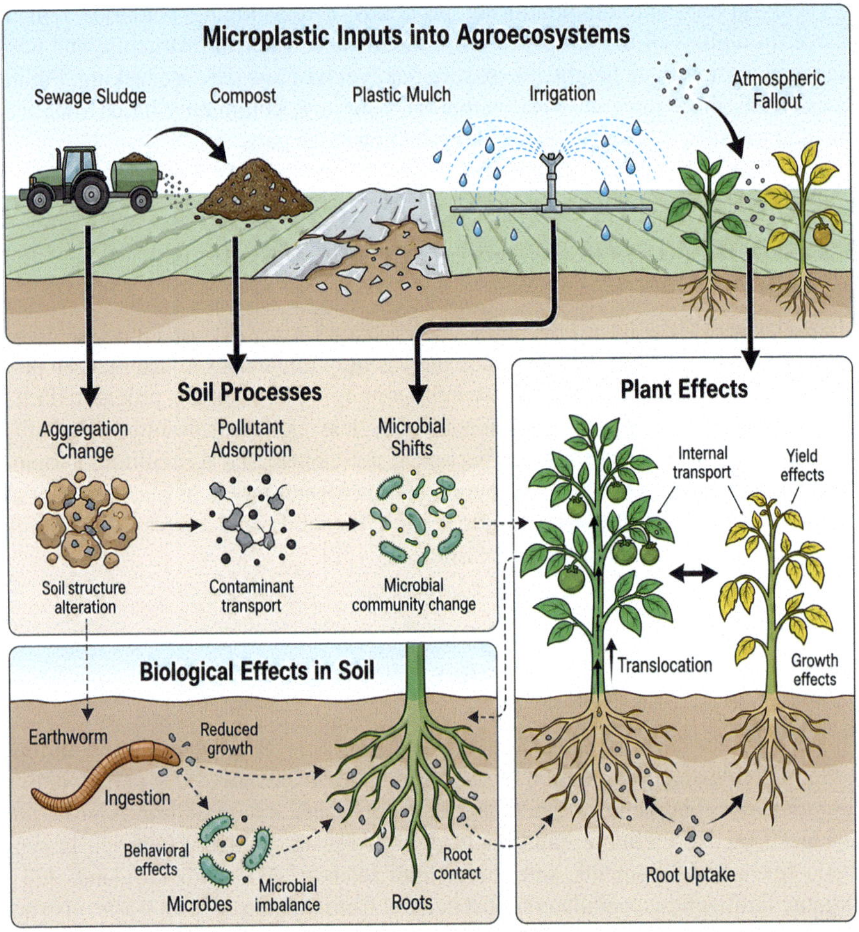

Fig. 5.2 Pathways and impacts of microplastics in agroecosystems. (Conceptual framework developed by the authors and illustrated using digital design tools)

microorganisms, leading to ingestion, behavioral changes, microbial imbalance, and potential reductions in organism growth. At the plant level, the schematic shows how microplastics may enter root systems through root contact and uptake, then translocate within plant tissues. These interactions may ultimately influence plant physiological performance, including internal transport processes, growth responses, and possible yield effects. Collectively, the figure presents an integrated view of how microplastic inputs, soil processes, biological interactions, and plant responses are interconnected, emphasizing the potential implications of plastic contamination for soil health and agricultural productivity.

Once introduced into the soil environment, microplastics may alter soil aggregation, adsorb organic and inorganic contaminants, and influence microbial community structure. These changes can affect soil fauna, such as earthworms and microorganisms, and may influence root interactions, plant uptake, and, ultimately, crop productivity.

5.2.2 *Effects on Soil Structure and Physicochemical Properties*

Microplastics alter soil physical structure by affecting porosity, bulk density, and water retention. de Souza Machado et al. [6] showed that different plastic types and shapes (e.g., fibers, films, and fragments) affect soil aggregation differently. For instance, film-like MPs reduce water evaporation rates and form preferential flow channels, which compromise the uniformity of water and nutrient distribution [38].

Furthermore, MPs can adsorb heavy metals and organic pollutants due to their high surface area and hydrophobic nature, acting as secondary vectors of contamination in soils [39]. These pollutants may desorb under changing pH or temperature conditions, affecting contaminant mobility and bioavailability in agricultural landscapes.

5.2.3 *Impacts on Soil Biota and Microbial Function*

Microplastics negatively affect soil biodiversity by altering the composition, abundance, and behavior of soil fauna, including earthworms, nematodes, and collembolans. Earthworms exposed to polyethylene microplastics showed reduced weight gain, cocoon production, and enzyme activity [40]. Moreover, MPs impair burrowing behavior and feeding efficiency, with cascading effects on nutrient cycling and organic matter decomposition.

At the microbial level, changes in community composition, enzymatic activity, and gene expression profiles have been observed. MPs can disrupt microbe-plant interactions, including nitrogen-fixing symbioses, by altering rhizosphere conditions [41]. Aralappanavar et al. [42] observed that plastic pollution altered microbial respiration rates and increased greenhouse gas emissions in contaminated soils.

5.2.4 *Plant Uptake and Translocation*

Recent studies have revealed that micro- and nanoplastics can be absorbed by crop roots and translocated to aerial tissues. Li et al. [43] demonstrated that lettuce and wheat roots internalize nanoplastics via endocytosis, and that subsequent accumulation in edible parts poses risks to human consumption. Further, such uptake may disrupt nutrient acquisition, photosynthesis, and overall plant health.

Importantly, MPs can also damage root ultrastructure and reduce root elongation, affecting plant vigor and yield. For instance, Li et al. [44] observed that cucumber plants exposed to polystyrene nanoparticles exhibited significant reductions in chlorophyll content, biomass, and seed germination rates.

5.2.5 *Threats to Agroecosystem Sustainability*

The persistence and ubiquity of MPs challenge sustainable farming practices. Soil health degradation, compromised crop quality, and reduced biodiversity affect long-term ecosystem services. Additionally, the socioeconomic implications for farmers, particularly those in low-resource settings who rely on compost and plasticulture technologies, warrant urgent attention.

Microplastics threaten the regulatory assumptions behind organic certification and sustainable labeling. As plastic residues become increasingly unavoidable in composts and biosolids, it becomes difficult to guarantee "plastic-free" agricultural produce [45].

5.2.6 *Knowledge Gaps and Future Directions*

Although awareness has grown, critical gaps remain in understanding MPs' long-term behavior in soil matrices. Most studies are short-term and lab-based, failing to replicate complex field conditions. Additionally, the interactions between microplastics and agrochemicals (e.g., pesticides and fertilizers) and their synergistic toxicities require comprehensive investigation.

Efforts must also address the detection challenges of MPs in soil, particularly in distinguishing synthetic polymers from organic residues. Advanced spectroscopic and thermal techniques, such as FTIR and Py-GC/MS, remain limited in detecting nanoplastics in complex matrices. Standardized protocols for monitoring, quantification, and risk assessment in agricultural contexts remain largely absent.

Research must increasingly focus on alternatives to plastic-based agricultural inputs, biodegradable mulch films, and policy mechanisms that regulate plastic use in food systems. Furthermore, inclusive knowledge-transfer strategies must be developed, particularly targeting smallholder farmers in the Global South.

5.3 Impacts on Biodiversity and Food Webs

Microplastics and nanoplastics have introduced unprecedented ecological stressors, destabilizing biodiversity across marine, freshwater, and terrestrial food webs. These synthetic particles occur across a wide size spectrum, from macroplastics to nanoplastics, and undergo continuous fragmentation through physical, chemical, biological, and photodegradation processes. Their persistence, mobility, and bioavailability enable them to disrupt species interactions, energy transfer, and trophic integrity. As illustrated in Fig. 5.3, the progressive breakdown of plastics increases their accessibility to lower trophic organisms, facilitating their entry into aquatic food webs and amplifying ecological risks across biological hierarchies. This

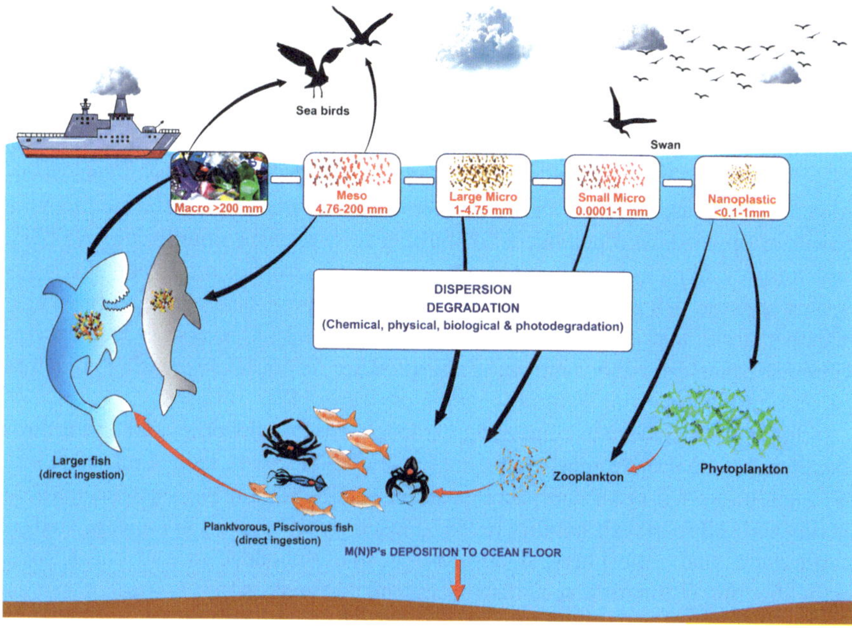

Fig. 5.3 Size-dependent dispersion, degradation, trophic transfer, and ecological impacts of microplastics and nanoplastics in aquatic food webs. (Adapted from [46]. This figure reproduced under the Creative Commons Attribution 4.0 International License (CC BY 4.0) (https://creativecommons.org/licenses/by/4.0/))

section synthesizes the multidimensional impacts of microplastics on biodiversity, emphasizing trophic cascades, species vulnerability, and ecosystem resilience.

The ingestion of microplastics by zooplankton, benthic invertebrates, and filter feeders compromises the energetic foundation of aquatic food webs by interfering with feeding processes and nutrient transfer [16, 47]. Experimental studies have demonstrated reduced feeding efficiency, reproductive impairment, and altered metabolic pathways in copepods and daphnids exposed to polystyrene particles [15, 48]. These sublethal effects can reduce population fitness at lower trophic levels and propagate ecological disruptions upward through the food web, ultimately affecting fish recruitment and predator success [21]. As illustrated in Fig. 5.3, the progressive fragmentation of plastic debris from macroplastics to mesoplastics, microplastics, and nanoplastics through physical, chemical, biological, and photodegradation processes increases their biological availability and facilitates their entry into aquatic food webs through phytoplankton and zooplankton. These particles are subsequently transferred to invertebrates, fish, and higher predators, such as seabirds, through trophic transfer and bioaccumulation. The figure also highlights vertical transport processes, including particle deposition to the ocean floor, demonstrating how plastic size influences environmental fate and trophic exposure. Together, these processes intensify ecological consequences across trophic levels, including energy loss, behavioral alterations, oxidative stress, reproductive impairment, and population-level effects, ultimately showing how microplastic contamination can amplify ecological risks and destabilize the structure and function of aquatic ecosystems.

In coral reef systems, microplastics not only interfere with photosymbiotic relationships but also promote pathogen colonization, exacerbating coral bleaching and tissue necrosis [49, 50]. Similar patterns of microbial dysbiosis have been documented in sponges, bivalves, and echinoderms, where plastic surfaces facilitate the growth of opportunistic pathogens, disturbing host-microbe equilibrium [25, 51]. Predator-prey dynamics shift under exposure to microplastics. Visual predators such as juvenile fish mistake buoyant fragments for prey, leading to gut blockage, oxidative stress, and impaired prey recognition [52, 53]. Conversely, prey species accumulate plastics without activating anti-predator responses, increasing predation risk and further distorting natural selection pressures [17].

In terrestrial ecosystems, emerging evidence from earthworms and soil arthropods indicates that microplastics disrupt burrowing behavior, reduce reproduction, and alter microbial community composition, with cascading effects on nutrient cycling and soil food web stability [6, 54]. Plastics entering soils via sewage sludge, plasticulture, and atmospheric deposition become embedded in root zones, indirectly affecting pollinators, seed dispersers, and detritivores [55].

Trophic magnification of plastic-associated pollutants, including polycyclic aromatic hydrocarbons (PAHs), heavy metals, and bisphenols, has been confirmed across various taxa [56]. Apex predators such as seabirds, marine mammals, and sharks accumulate microplastics both directly through ingestion and indirectly via prey transfer, leading to chronic inflammation, endocrine disruption, and bioenergetic imbalances [57, 58].

Genetic and epigenetic consequences are now under investigation. Zebrafish (*Danio rerio*) and medaka models have revealed microplastic-induced transgenerational effects, such as altered gene expression in stress and detoxification pathways [59, 60]. These molecular perturbations threaten species resilience and adaptive capacity under climate change scenarios.

The structural integrity of food webs collapses when foundational species decline. Microplastic exposure undermines keystone species, such as mussels, amphipods, and krill, disrupting their ecological functions and impairing nutrient cycling and habitat provisioning [23, 61]. These losses reverberate across trophic networks, amplifying the risk of ecosystem regime shifts.

Biodiversity loss triggered by microplastics challenges conservation efforts. Protected areas, including marine sanctuaries and biodiversity hotspots, remain vulnerable to airborne and ocean-borne plastic inputs, undermining gains in restoration and ecological baselines [7]. As plastic production continues to rise, mitigating its ecological impacts requires coordinated research, policy, and community interventions to protect food web stability and biodiversity integrity.

5.4 Link to Food Security in the Global South

Microplastic pollution presents an increasingly complex threat to food security in the Global South, where communities often rely directly on local marine, freshwater, and agricultural resources for sustenance. These regions face overlapping challenges, including resource vulnerability, high dependence on subsistence farming and artisanal fisheries, limited waste infrastructure, and exposure to imported plastic-containing goods. The persistent intrusion of microplastics into ecological systems complicates these pressures by affecting food quantity, quality, and safety.

Artisanal and small-scale fisheries, which supply a major share of protein intake for coastal populations in Southeast Asia and Sub-Saharan Africa, are increasingly vulnerable to declining fish health and reproductive capacity associated with microplastic ingestion and exposure to plastic-bound contaminants [18, 59, 60]. The contamination of seafood with microplastics, therefore, represents an underrecognized dietary exposure pathway, particularly for low-income communities with limited dietary diversity and a high dependence on aquatic resources [61]. In freshwater environments, benthic feeders such as tilapia, widely cultivated in aquaculture systems across the Global South, have been shown to accumulate microplastic fibers and fragments that can impair growth, metabolism, and fecundity [38, 62].

As illustrated in Fig. 5.4, microplastic contamination originates from interconnected environmental pathways, including water contamination from river discharge and coastal pollution, soil contamination from agricultural exposure, and atmospheric deposition through plastic fallout. These contamination routes converge within food systems through seafood exposure and agricultural contamination, creating multiple entry points into human diets. The framework further shows how human exposure occurs through dietary intake and drinking water, with chronic exposure risks particularly affecting vulnerable populations in the Global South.

The figure also emphasizes how structural factors shape these risks. Waste infrastructure limitations, such as open dumping, landfill overflow, and river leakage, act as upstream drivers of environmental contamination. At the same time, monitoring and data gaps, including fragmented monitoring systems, data uncertainty, and a lack of standardized methods, limit effective risk assessment and management. These challenges are further compounded by socioeconomic vulnerabilities that affect smallholder farmers, coastal communities, and informal waste sector workers, whose livelihoods and food security depend directly on environmental quality.

The framework, therefore, highlights the importance of policy and management responses to reduce these risks, including strengthening waste management systems, expanding environmental monitoring programs, improving food safety regulations, and promoting circular economy strategies. Together, the figure presents an integrated systems perspective demonstrating how environmental contamination, food system exposure, human health risks, and governance conditions interact to shape food security outcomes in microplastic-affected regions of the Global South.

In terrestrial food systems, irrigation using contaminated surface water and plastic mulching practices contribute to microplastic accumulation in agricultural soils

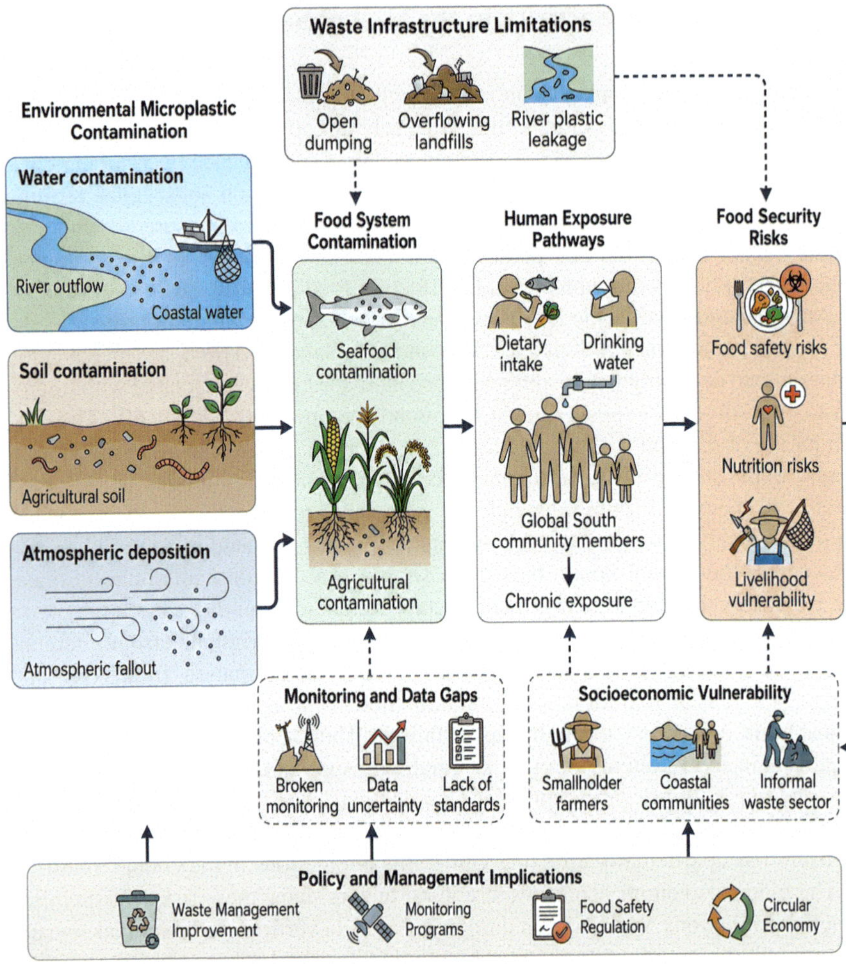

Fig. 5.4 Microplastics-ecosystem-food security nexus in the Global South. (Conceptual framework developed by the authors and illustrated using digital design tools)

[37]. Uptake of nanoplastics by crops such as rice, lettuce, and wheat—staples across Asia and Africa—raises concern for trophic transfer within human food chains [42, 63]. The adsorption of heavy metals, pathogens, and agrochemical residues onto microplastic surfaces amplifies these risks by introducing secondary toxicants [64].

Food security in the Global South is further undermined by limited monitoring, the absence of regulatory frameworks, and constrained research infrastructure for microplastic surveillance [65]. Socioeconomic vulnerability exacerbates exposure and reduces adaptive capacity, creating disproportionate health and development burdens. Efforts to address these challenges must integrate microplastics into broader strategies for food system resilience, public health policy, and environmental governance across developing regions.

5.5 Chapter Synthesis

This chapter shows that microplastics function as ecological stressors that influence biological systems from the cellular level to whole ecosystems. Evidence across marine, freshwater, and terrestrial environments confirms that these particles alter organism physiology, disrupt ecological interactions, and weaken ecosystem stability. Their effects extend beyond physical presence. Their role as carriers of chemical contaminants and microbial communities increases ecological risk and complicates predictions of long-term environmental consequences.

The discussion of aquatic environments demonstrates that microplastics interfere with feeding processes, reproductive success, and metabolic performance across multiple trophic levels. Their ingestion by plankton, invertebrates, and fish introduces energetic costs that reduce organism fitness and alter predator-prey relationships. Sediment accumulation further affects benthic processes, nutrient exchange, and habitat quality. These changes influence ecosystem productivity and resilience, particularly in highly stressed coastal and estuarine environments.

This chapter also establishes that terrestrial ecosystems face comparable risks. Microplastic accumulation in agricultural soils affects soil structure, microbial processes, and plant development. Their interactions with fertilizers, pesticides, and heavy metals introduce complex exposure pathways that extend from soil systems into food production. These processes raise important concerns for long-term soil sustainability and agricultural productivity.

The analysis of biodiversity impacts further demonstrates that microplastics influence ecological stability through trophic disruption and species-level stress. Sublethal biological effects accumulate across food webs and may reduce ecosystem resilience under additional environmental pressures, such as climate change and habitat degradation. The persistence of microplastics therefore represents a chronic ecological pressure rather than a short-term disturbance.

This chapter also highlights the intersection between ecological integrity and food security. Regions that depend heavily on fisheries and smallholder agriculture face elevated exposure risks due to limited waste management capacity and high environmental dependence. These findings reinforce the need to integrate ecological science with food system governance and environmental protection strategies.

The evidence presented in this chapter confirms that microplastic pollution represents a systemic ecological challenge that affects environmental stability and biological resilience. Addressing this challenge requires advances in ecological monitoring, pollution prevention strategies, and sustainable material management. A strong scientific understanding of ecological impacts, therefore, serves as an essential foundation for evaluating the broader environmental and societal consequences of plastic pollution.

Understanding ecological impacts clarifies how microplastics affect environmental systems and food production networks. The next chapter extends this discussion by examining how these environmental exposures translate into human exposure pathways. Attention shifts toward the presence of microplastics in food, water, and air, their potential toxicological implications, and the emerging evidence linking environmental contamination with human health risks.

References

1. Rezania, S., Park, J., Md Din, M. F., Mat Taib, S., Talaiekhozani, A., Kumar Yadav, K., & Kamyab, H. (2018). Microplastics pollution in different aquatic environments and biota: A review of recent studies. *Marine Pollution Bulletin, 133*, 191–208. https://doi.org/10.1016/j.marpolbul.2018.05.022
2. Hirt, N., & Body-Malapel, M. (2020). Immunotoxicity and intestinal effects of nano- and microplastics: A review of the literature. *Particle and Fibre Toxicology, 17*, 57. https://doi.org/10.1186/s12989-020-00387-7
3. Hossain, M., & Engelhardt, I. (2025). Global plastic footprint: Unveiling property trends, environmental fate, and emerging threats of microplastic and nanoplastics pollution across ecosystems. *Energy Ecology and Environment, 10*(6), 637. https://doi.org/10.1007/s40974-025-00383-7
4. Akdogan, Z., & Guven, B. (2019). Microplastics in the environment: A critical review of current understanding and identification of future research needs. *Environmental Pollution, 254*, 113011. https://doi.org/10.1016/j.envpol.2019.113011
5. Akindele, E. O., Ehlers, S. M., & Koop, J. H. E. (2020). Freshwater insects of different feeding guilds ingest microplastics in two Gulf of Guinea tributaries in Nigeria. *Environmental Science and Pollution Research, 27*, 33373–33379. https://doi.org/10.1007/s11356-020-08763-8
6. de Souza Machado, A. A., Lau, C. W., Till, J., Kloas, W., Lehmann, A., Becker, R., & Rillig, M. C. (2018). Impacts of microplastics on the soil biophysical environment. *Environmental Science & Technology, 52*(17), 9656–9665. https://doi.org/10.1021/acs.est.8b02212
7. Galgani, F., Hanke, G., Werner, S., & De Vrees, L. (2013). Marine litter within the European Marine Strategy Framework Directive. *ICES Journal of Marine Science, 70*(6), 1055–1064. https://doi.org/10.1093/icesjms/fst122
8. Barrows, A. P., Neumann, C. A., Berger, M. L., & Shaw, S. D. (2017). Grab vs. neuston tow net: A microplastic sampling performance comparison and possible advances in the field. *Analytical Methods, 9*, 1446–1453. https://doi.org/10.1039/C6AY02387H
9. Dris, R., Gasperi, J., Saad, M., Mirande, C., & Tassin, B. (2016). Synthetic fibers in atmospheric fallout: A source of microplastics in the environment? *Marine Pollution Bulletin, 104*(1–2), 290–293. https://doi.org/10.1016/j.marpolbul.2016.01.006
10. Eerkes-Medrano, D., Thompson, R. C., & Aldridge, D. C. (2015). Microplastics in freshwater systems: A review of the emerging threats, identification of knowledge gaps and prioritisation of research needs. *Water Research, 75*, 63–82. https://doi.org/10.1016/j.watres.2015.02.012
11. Chaukura, N., Kefeni, K. K., Chikurunhe, I., Nyambiya, I., Gwenzi, W., Moyo, W., Nkambule, T. I. I., Mamba, B. B., & Abulude, F. O. (2021). Microplastics in the aquatic environment: The occurrence, sources, ecological impacts, fate, and remediation challenges. *Pollutants, 1*(2), 95–118. https://doi.org/10.3390/pollutants1020009
12. Lin, H.-T., Schneider, F., Aziz, M. A., Wong, K. Y., Arunachalam, K. D., Praveena, S. M., Sethupathi, S., Chong, W. C., Nafisyah, A. L., Parthasarathy, P., Chelliapan, S., & Kunz, A. (2024). Microplastics in Asian rivers: Geographical distribution, most detected types, and inconsistency in methodologies. *Environmental Pollution, 349*, 123985. https://doi.org/10.1016/j.envpol.2024.123985
13. Strady, E., Dang, T. H., Dao, T. D., Dinh, H. N., Dung, T. T., Duong, T. N., Duong, T. T., Hoang, D. A., Kieu-Le, T.-C., Le, T. P. Q., Mai, H., Trinh-Dang, M., Nguyen, Q. H., Tran-Nguyen, Q. A., Tran, Q. V., Truong, T. N. S., Chu, V. H., & Vo, V. C. (2020). Baseline assessment of microplastic concentrations in marine and freshwater environments of a

developing Southeast Asian country, Viet Nam. *Marine Pollution Bulletin, 162*, 111870. https://doi.org/10.1016/j.marpolbul.2020.111870
14. Cole, M., Lindeque, P., Fileman, E., Halsband, C., & Galloway, T. S. (2015). The impact of polystyrene microplastics on feeding, function, and fecundity in the marine copepod *Calanus helgolandicus*. *Environmental Science & Technology, 49*(2), 1130–1137. https://doi.org/10.1021/es504525u
15. Besseling, E., Wang, B., Lürling, M., & Koelmans, A. A. (2014). Nanoplastic affects growth of *S. obliquus* and reproduction of *D. magna*. *Environmental Science & Technology, 48*(20), 12336–12343. https://doi.org/10.1021/es503001d
16. Setälä, O., Fleming-Lehtinen, V., & Lehtiniemi, M. (2014). Ingestion and transfer of microplastics in the planktonic food web. *Environmental Pollution, 185*, 77–83. https://doi.org/10.1016/j.envpol.2013.10.013
17. Wright, S. L., Rowe, D., Thompson, R. C., & Galloway, T. S. (2013). Microplastic ingestion decreases energy reserves in marine worms. *Current Biology, 23*(23), PR1031–R1033. https://doi.org/10.1016/j.cub.2013.10.068
18. Tang, K. H. D., Li, R., Li, Z., & Wang, D. (2024). Health risk of human exposure to microplastics: A review. *Environmental Chemistry Letters, 22*(3), 1155. https://doi.org/10.1007/s10311-024-01727-1
19. Rochman, C. M., Hoh, E., Hentschel, B. T., & Kaye, S. (2013). Long-term field measurement of sorption of organic contaminants to five types of plastic pellets: Implications for plastic marine debris. *Environmental Science & Technology, 47*(3), 1646–1654. https://doi.org/10.1021/es303700s
20. Bandee, M., Multisanti, C. R., Impellitteri, F., Piccione, G., & Faggio, C. (2025). Environmental toxicology of microplastic particles on fish: A review. *Comparative Biochemistry and Physiology Part C: Toxicology & Pharmacology, 287*, 110042. https://doi.org/10.1016/j.cbpc.2024.110042
21. Mazurais, D., Ernande, B., Quazuguel, P., Severe, A., Huelvan, C., Madec, L., Mouchel, O., Soudant, P., Robbens, J., Huvet, A., & Zambonino-Infante, J. (2015). Evaluation of the impact of polyethylene microbeads ingestion in European sea bass (*Dicentrarchus labrax*) larvae. *Marine Environmental Research, 112*(Part A), 78–85. https://doi.org/10.1016/j.marenvres.2015.09.009
22. von Moos, N., Burkhardt-Holm, P., & Köhler, A. (2012). Uptake and effects of microplastics on cells and tissue of the blue mussel *Mytilus edulis* L. after an experimental exposure. *Environmental Science & Technology, 46*(20), 11327–11335. https://doi.org/10.1021/es302332w
23. Farrell, P., & Nelson, K. (2013). Trophic level transfer of microplastic: *Mytilus edulis* (L.) to *Carcinus maenas* (L.). *Environmental Pollution, 177*, 1–3. https://doi.org/10.1016/j.envpol.2013.01.046
24. Lusher, A. L., McHugh, M., & Thompson, R. C. (2013). Occurrence of microplastics in the gastrointestinal tract of pelagic and demersal fish from the English Channel. *Marine Pollution Bulletin, 67*(1–2), 94–99. https://doi.org/10.1016/j.marpolbul.2012.11.028
25. Caruso, G. (2015). Plastic degrading microorganisms as a tool for bioremediation of plastic contamination in aquatic environments. *Journal of Pollution Effects & Control, 3*(3), e112. https://doi.org/10.4172/2375-4397.1000e112
26. Pal, D., Prabhakar, R., Barua, V. B., Zekker, I., Burlakovs, J., Krauklis, A. E., Hogland, W., & Vincēviča–Gaile, Z. (2024). Microplastics in aquatic systems: A comprehensive review of its distribution, environmental interactions, and health risks. *Environmental Science and Pollution Research, 32*(1), 56. https://doi.org/10.1007/s11356-024-35741-1
27. Seeley, M. E., Song, B., Passie, R., & Hale, R. C. (2020). Microplastics affect sedimentary microbial communities and nitrogen cycling. *Nature Communications, 11*(1), 2372. https://doi.org/10.1038/s41467-020-16235-3
28. Green, D. S., Boots, B., Blockley, D. J., Rocha, C., & Thompson, R. (2015). Impacts of discarded plastic bags on marine assemblages and ecosystem functioning. *Environmental Science & Technology, 49*(9), 5380–5389. https://doi.org/10.1021/acs.est.5b00277
29. Mani, T., Hauk, A., Walter, U., & Burkhardt-Holm, P. (2015). Microplastics profile along the Rhine River. *Scientific Reports, 5*, 17988. https://doi.org/10.1038/srep17988

30. Swain, P. R., Parida, P. K., Majhi, P., Behera, B. K., & Das, B. K. (2025). Microplastics as emerging contaminants: Challenges in inland aquatic food web. *Water, 17*(2), 201. https://doi.org/10.3390/w17020201
31. Bläsing, M., & Amelung, W. (2018). Plastics in soil: Analytical methods and possible sources. *Science of the Total Environment, 612*, 422–435. https://doi.org/10.1016/j.scitotenv.2017.08.086
32. Corradini, F., Meza, P., Eguiluz, R., Casado, F., Huerta-Lwanga, E., & Geissen, V. (2019). Evidence of microplastic accumulation in agricultural soils from sewage sludge and plastic mulch. *Science of the Total Environment, 671*, 411–420. https://doi.org/10.1016/j.scitotenv.2019.03.368
33. Ramanayaka, S., Zhang, H., & Semple, K. T. (2024). Environmental fate of microplastics and common polymer additives in non-biodegradable plastic mulch applied agricultural soils. *Environmental Pollution, 363*, 125249. https://doi.org/10.1016/j.envpol.2024.125249
34. Nizzetto, L., Futter, M., & Langaas, S. (2016). Are agricultural soils dumps for microplastics of urban origin? *Environmental Science & Technology, 50*(20), 10777–10779. https://doi.org/10.1021/acs.est.6b04140
35. Weithmann, N., Möller, J. N., Löder, M. G. J., Piehl, S., Laforsch, C., & Freitag, R. (2018). Organic fertilizers as a vehicle for the entry of microplastic into the environment. *Science Advances, 4*(4), eaap8060. https://doi.org/10.1126/sciadv.aap8060
36. Dris, R., Gasperi, J., Rocher, V., Saad, M., Renault, N., & Tassin, B. (2015). Microplastic contamination in an urban area: A case study in Greater Paris. *Environmental Chemistry, 12*(5), 592–599. https://doi.org/10.1071/EN14167
37. Aydın, S., Mucevher, O., Ulvi, A., Bedük, F., Aydın, M. E., Merken, Ö., & Uzun, C. (2025). Effects of long-term wastewater irrigation on microplastics pollution in agricultural soil. *Environmental Science and Pollution Research, 32*(20), 12340. https://doi.org/10.1007/s11356-025-36452-x
38. Zhang, G. S., & Liu, Y. F. (2018). The distribution of microplastics in soil aggregate fractions in southwestern China. *Science of the Total Environment, 642*, 12–20. https://doi.org/10.1016/j.scitotenv.2018.06.004
39. Wang, J., Liu, X., Li, Y., Powell, T., Wang, X., Wang, G., & Zhang, P. (2019). Microplastics as contaminants in the soil environment: A mini-review. *Science of the Total Environment, 691*, 848–857. https://doi.org/10.1016/j.scitotenv.2019.07.209
40. Huerta Lwanga, E., Gertsen, H., Gooren, H., Peters, P., Salánki, T., van der Ploeg, M., Besseling, E., Koelmans, A., & Geissen, V. (2016). Microplastics in the terrestrial ecosystem: Implications for *Lumbricus terrestris* (Oligochaeta, Lumbricidae). *Environmental Science & Technology, 50*(5), 2685–2691. https://doi.org/10.1021/acs.est.5b05478
41. Zhai, Y., Bai, J., Chang, P., Liu, Z., Wang, Y., Liu, G., Cui, B., Peijnenburg, W. J. G. M., & Vijver, M. G. (2024). Microplastics in terrestrial ecosystem: Exploring the menace to the soil-plant-microbe interactions. *TrAC Trends in Analytical Chemistry, 174*, 117667. https://doi.org/10.1016/j.trac.2024.117667
42. Aralappanavar, V. K., Mukhopadhyay, R., Yu, Y., Liu, J., Bhatnagar, A., Praveena, S. M., Li, Y., Paller, M., Aydel, T. M., Rinklebe, J., Bolan, N. S., & Sarkar, B. (2024). Effects of microplastics on soil microorganisms and microbial functions in nutrients and carbon cycling—A review. *Science of the Total Environment, 924*, 171435. https://doi.org/10.1016/j.scitotenv.2024.171435
43. Li, L., Luo, Y., Li, R., Zhou, Q., Peijnenburg, W. J. G. M., Yin, N., Yang, J., Tu, C., & Zhang, Y. (2020). Effective uptake of submicrometre plastics by crop plants via a crack-entry mode. *Nature Sustainability, 3*(11), 929–937. https://doi.org/10.1038/s41893-020-0567-9
44. Li, Z., Li, R., Li, Q., Zhou, J., & Wang, G. (2020). Physiological response of cucumber (*Cucumis sativus* L.) leaves to polystyrene nanoplastics pollution. *Chemosphere, 255*, 127041. https://doi.org/10.1016/j.chemosphere.2020.127041
45. Büks, F., & Kaupenjohann, M. (2020). Global concentrations of microplastics in soils—A review. *The Soil, 6*(2), 649–662. https://doi.org/10.5194/soil-6-649-2020
46. Benson, N. U., Agboola, O. D., Fred-Ahmadu, O. H., De-la-Torre, G. E., Oluwalana, A., & Williams, A. B. (2022). Micro(nano)plastics prevalence, food web interactions, and toxicity assessment in aquatic organisms: A review. *Frontiers in Marine Science, 9*, 851281. https://doi.org/10.3389/fmars.2022.851281

47. Cole, M., Lindeque, P., Halsband, C., & Galloway, T. S. (2011). Microplastics as contaminants in the marine environment: A review. *Marine Pollution Bulletin, 62*(12), 2588–2597. https://doi.org/10.1016/j.marpolbul.2011.09.025
48. Rist, S., Baun, A., & Hartmann, N. B. (2017). Ingestion of micro- and nanoplastics in *Daphnia magna*—Quantification of body burdens and assessment of feeding rates and reproduction. *Environmental Pollution, 228*, 398–407. https://doi.org/10.1016/j.envpol.2017.05.048
49. Hall, N. M., Berry, K. L. E., Rintoul, L., & Hoogenboom, M. O. (2015). Microplastic ingestion by scleractinian corals. *Marine Biology, 162*, 725–732. https://doi.org/10.1007/s00227-015-2619-7
50. Lamb, J. B., Willis, B. L., Fiorenza, E. A., Couch, C. S., Howard, R., Rader, D. N., True, J. D., Kelly, L. A., Ahmad, A., Jompa, J., & Harvell, C. D. (2018). Plastic waste associated with disease on coral reefs. *Science, 359*(6374), 460–462. https://doi.org/10.1126/science.aar3320
51. Kirstein, I. V., Wichels, A., Krohne, G., & Gerdts, G. (2018). Mature biofilm communities on synthetic polymers in seawater—Specific or general? *Marine Environmental Research, 142*, 147–154. https://doi.org/10.1016/j.marenvres.2018.09.028
52. Roch, S., Friedrich, C., & Brinker, A. (2020). Uptake routes of microplastics in fishes: Practical and theoretical approaches to test existing theories. *Scientific Reports, 10*, 3896. https://doi.org/10.1038/s41598-020-60630-1
53. Critchell, K., & Hoogenboom, M. O. (2018). Effects of microplastic exposure on the body condition and behaviour of planktivorous reef fish (*Acanthochromis polyacanthus*). *PLoS One, 13*(3), e0193308. https://doi.org/10.1371/journal.pone.0193308
54. Paudel, P., Kumar, R., Pandey, M. K., Paudel, P., & Subedi, M. (2024). Exploring the impact of micro-plastics on soil health and ecosystem dynamics: A comprehensive review. *Journal of Experimental Biology and Agricultural Sciences, 12*(2), 163. https://doi.org/10.18006/2024.12(2).163.174
55. Chae, Y., & An, Y. J. (2018). Current research trends on plastic pollution and ecological impacts on the soil ecosystem: A review. *Environmental Pollution, 240*, 387–395. https://doi.org/10.1016/j.envpol.2018.05.008
56. Koelmans, A. A., Bakir, A., Burton, G. A., & Janssen, C. R. (2016). Microplastic as a vector for chemicals in the aquatic environment: Critical review and model-supported reinterpretation of empirical studies. *Environmental Science & Technology, 50*(7), 3315–3326. https://doi.org/10.1021/acs.est.5b06069
57. Wilcox, C., Van Sebille, E., & Hardesty, B. D. (2015). Threat of plastic pollution to seabirds is global, pervasive, and increasing. *Proceedings of the National Academy of Sciences, 112*(38), 11899–11904. https://doi.org/10.1073/pnas.1502108112
58. Nelms, S. E., Barnett, J., Brownlow, A., Davison, N. J., Deaville, R., Galloway, T. S., Lindeque, P. K., Santillo, D., & Godley, B. J. (2019). Microplastics in marine mammals stranded around the British coast: Ubiquitous but transitory? *Scientific Reports, 9*, 1075. https://doi.org/10.1038/s41598-018-37428-3
59. Jin, Y., Xia, J., Pan, Z., Yang, J., Wang, W., & Fu, Z. (2018). Polystyrene microplastics induce microbiota dysbiosis and inflammation in the gut of adult zebrafish. *Environmental Pollution, 235*, 322–329. https://doi.org/10.1016/j.envpol.2017.12.088
60. Pitt, J. A., Kozal, J. S., Jayasundara, N., Massarsky, A., Trevisan, R., Geitner, N., Wiesner, M., Levin, E. D., & Di Giulio, R. T. (2018). Uptake, tissue distribution, and toxicity of polystyrene nanoparticles in developing zebrafish (*Danio rerio*). *Aquatic Toxicology, 194*, 185–194. https://doi.org/10.1016/j.aquatox.2017.11.017
61. Browne, M. A., Dissanayake, A., Galloway, T. S., Lowe, D. M., & Thompson, R. C. (2008). Ingested microscopic plastic translocates to the circulatory system of the mussel, Mytilus edulis. *Environmental Science & Technology, 42*(13), 5026–5031. https://doi.org/10.1021/es800249a
62. Barboza, L. G. A., Vethaak, A. D., Lavorante, B. R. B. O., Lundebye, A. K., & Guilhermino, L. (2018). Marine microplastic debris: An emerging issue for food security, food safety and human health. *Marine Pollution Bulletin., 133*, 336–348. https://doi.org/10.1016/j.marpolbul.2018.05.047

63. Smith, M., Love, D. C., Rochman, C. M., & Neff, R. A. (2018). Microplastics in seafood and the implications for human health. *Current Environmental Health Reports, 5*(3), 375–386. https://doi.org/10.1007/s40572-018-0206-z
64. Jabeen, K., Su, L., Li, J., Yang, D., Tong, C., Mu, J., & Shi, H. (2017). Microplastics and mesoplastics in fish from coastal and fresh waters of China. *Environmental Pollution, 221*, 141–149. https://doi.org/10.1016/j.envpol.2016.11.055
65. Rehman, A., Habumugisha, T., Huang, F., Zhang, Z., Kiki, C., Abdullah, M. A., Yan, C., Shaheen, U., & Zhang, X. (2025). Impacts of polystyrene nanoplastics on zebrafish gut microbiota and mechanistic insights. *Ecotoxicology and Environmental Safety, 299*, 118332. https://doi.org/10.1016/j.ecoenv.2025.118332

Chapter 6
Microplastics and Human Health

Introduction

Microplastics have shifted from being an environmental concern to a direct human health issue, following their consistent detection in human tissues and fluids. Recent evidence confirms the presence of microplastics in blood, lungs, placenta, breast milk, semen, and internal organs, indicating that exposure is continuous and systemic rather than incidental [1]. Humans encounter microplastics through multiple everyday pathways, including contaminated food and drinking water, inhalation of airborne fibers, and contact with consumer and occupational products. These exposure routes operate across all life stages, from prenatal development to adulthood, raising concerns about long-term health consequences.

Toxicological studies demonstrate that microplastics can induce oxidative stress, inflammation, endocrine disruption, and cellular damage. Their small size facilitates tissue penetration, while their surfaces act as carriers for plastic additives, persistent organic pollutants, metals, and antibiotic-resistant microorganisms. This dual role as particles and chemical vectors complicates health risk assessment, as biological effects depend on polymer type, size, shape, concentration, and associated chemicals. Experimental and emerging clinical evidence links microplastic exposure to gastrointestinal disorders, respiratory inflammation, metabolic dysfunction, cardiovascular stress, and potential reproductive effects, although causal pathways remain under active investigation [1].

Despite rapid advances, significant gaps persist in exposure quantification, standardized detection methods, and epidemiological evidence. Addressing these gaps is essential for translating scientific findings into effective public health policy and regulatory frameworks, particularly as global plastic production continues to rise.

R. Y. Capangpangan, A. C. Alguno, *Microplastics in Focus*,
https://doi.org/10.1007/978-3-032-28978-0_6

6.1 Pathways to Humans: Food, Water, Air, and Occupational Exposure

Microplastics infiltrate the human body through multiple, interconnected exposure routes, including ingestion through contaminated food and drinking water, inhalation of airborne particles, and dermal exposure through consumer products and contact with contaminated water in occupational and daily environments [2, 3]. The widespread presence of microplastics across environmental compartments results in continuous, multi-route human exposure from early life stages through adulthood.

As illustrated in Fig. 6.1, these three principal exposure routes originate from diverse sources, including urban and indoor air, industrial emissions, contaminated food and drinking water, food packaging materials, personal care products, contaminated recreational waters, and medical materials. The figure further shows how these pathways are associated with specific target organs, including the lungs and systemic circulation via inhalation, the gastrointestinal tract and liver via ingestion, and the skin and subcutaneous tissues via dermal absorption. It also highlights associated health impacts, such as inflammation, oxidative stress, respiratory dysfunction, microbiome alterations, and potential systemic distribution, emphasizing the cumulative nature of multi-route exposure and the potential human health risks posed by microplastics.

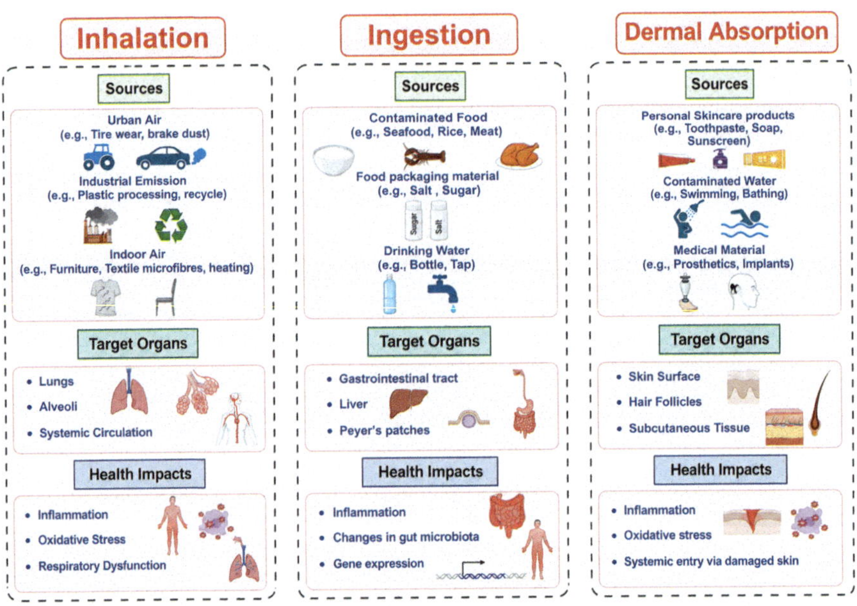

Fig. 6.1 Major human exposure pathways, target organs, and health impacts associated with microplastic exposure. (Adapted from [4]. This figure reproduced under the Creative Commons Attribution 4.0 International License (CC BY 4.0) (https://creativecommons.org/licenses/by/4.0/))

6.1.1 Ingestion Via Food and Water

Human exposure through ingestion primarily arises from the consumption of seafood, salt, honey, fruits, vegetables, and processed foods [5, 6]. Marine organisms, including fish and shellfish, accumulate microplastics in their tissues, making seafood a significant vector [7]. A study by Li et al. [8] confirmed microplastics in the gastrointestinal tracts of commercially available fish. Moreover, bottled and tap water worldwide contain detectable microplastic particles, with reported concentrations ranging from 0.1 to over 300 particles per liter [9, 10]. Tea bags, plastic packaging, and infant feeding bottles also contribute to dietary exposure [11].

6.1.2 Inhalation of Airborne Microplastics

Airborne microplastics pose a substantial inhalation risk, especially in indoor environments with synthetic textiles, carpets, and household dust [12]. Atmospheric fallout studies have identified microplastics in both urban and remote environments, with indoor air concentrations significantly higher due to limited ventilation and synthetic furnishings [13]. The inhaled particles may deposit in the respiratory tract, potentially triggering oxidative stress and inflammation [14]. Occupational settings such as plastic manufacturing plants, textile factories, and recycling facilities present even higher exposure levels for workers [15].

6.1.3 Occupational Exposure

Workers in industries that process plastic, produce synthetic fibers, and manage waste are exposed to higher levels of airborne and dermal microplastics [16]. Repetitive exposure in confined settings without adequate protective measures can result in respiratory irritation, dermatoses, and possible systemic effects [17]. Despite this elevated risk, current occupational safety standards do not specifically regulate exposure to microplastic particles.

6.1.4 Combined and Cumulative Risks

Pathways to human exposure do not operate in isolation. Multiple routes often interact, leading to cumulative and possibly synergistic effects [18]. For instance, an individual may ingest microplastics from seafood, inhale fibers from indoor air, and experience dermal contact at the workplace. The combined exposure raises concerns about bioaccumulation and long-term health impacts. Vulnerable populations, including pregnant women, infants, and frontline workers, require urgent attention due to their higher susceptibility [19].

6.2 Potential Health Risks: Toxicity, Additives, and Bioaccumulation

Microplastics pose complex health risks due to their dual nature as physical particles and chemical carriers. Once internalized through ingestion, inhalation, or dermal exposure, these particles can interact with cells and tissues at multiple biological levels. Their small size, particularly in the nanometer range, allows them to penetrate biological barriers, such as the intestinal epithelium, pulmonary alveoli, and compromised skin layers, enabling entry into systemic circulation [20]. This ability raises concerns about long-term persistence, systemic distribution, and chronic exposure even at low concentrations. As illustrated in Fig. 6.2, microplastics and nanoplastics may enter the body through contaminated food and beverages, airborne particles, and skin contact with consumer products; they then translocate across biological barriers and potentially distribute to multiple organ systems. These processes may affect the brain, lungs, liver, gastrointestinal tract, circulatory system, reproductive organs, placenta, and immune system through mechanisms, such as inflammation, oxidative stress, endocrine disruption, altered microbiome, and cellular damage.

The physical properties of microplastics—including their shape, size, and polymer composition—contribute to their toxicological behavior. Irregularly shaped fibers and fragments have shown greater potential to cause mechanical stress, inflammation, and cellular membrane disruption in mammalian models [22]. Additionally, microplastics have been found to generate reactive oxygen species (ROS), promoting oxidative stress and mitochondrial dysfunction, which can lead to genotoxicity and immune dysregulation [2].

However, particle toxicity alone does not fully explain the observed effects. Microplastics also serve as vectors for a wide array of toxicants. These include legacy persistent organic pollutants (POPs), such as PCBs and DDT, heavy metals, such as lead and cadmium, and newly recognized endocrine-disrupting additives, such as bisphenol A (BPA), phthalates, and flame retardants [23, 24]. These substances, whether incorporated during manufacturing or adsorbed from the environment, can leach out once internalized, enter systemic circulation, and affect endocrine, metabolic, and reproductive systems.

One emerging concern is the bioaccumulation of these plastic-associated chemicals in fatty tissues, where they may persist and exert prolonged effects. Repeated exposure through food, water, and air—combined with the long half-life of many additives—can lead to cumulative toxicity. Epidemiological evidence linking high phthalate levels to insulin resistance, altered lipid profiles, and reproductive toxicity supports these concerns [25].

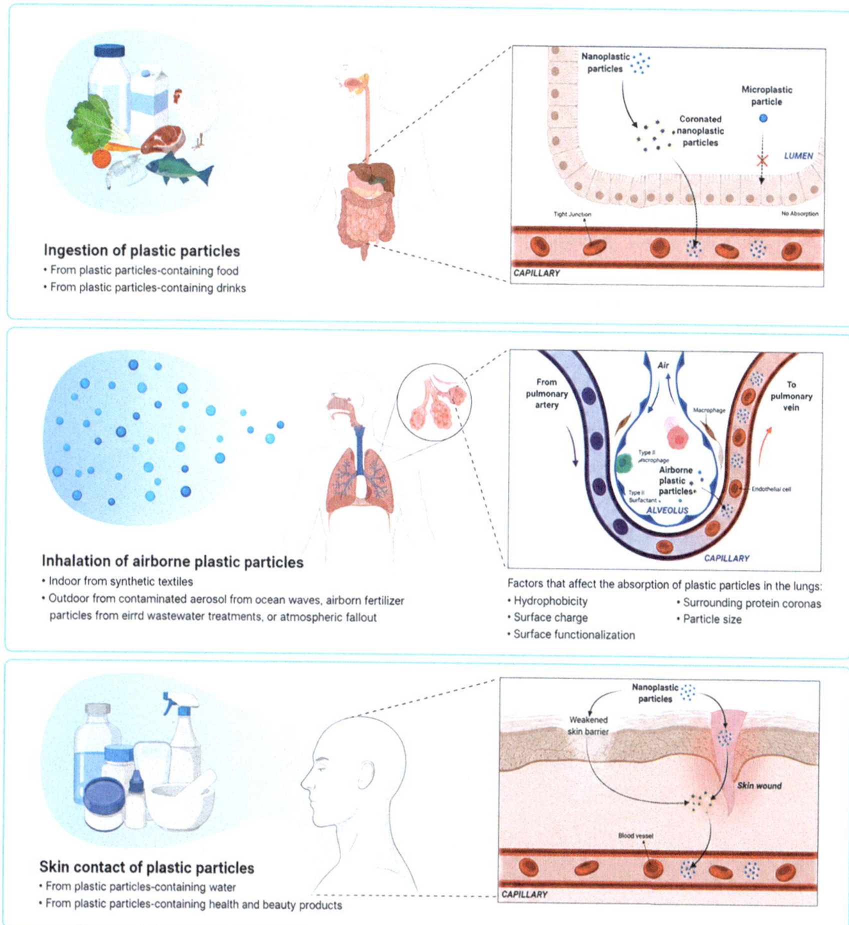

Fig. 6.2 Human exposure routes, biological uptake mechanisms, and systemic distribution pathways of microplastics and nanoplastics. (Adapted from [21]. This figure reproduced under the Creative Commons Attribution 4.0 International License (CC BY 4.0) (https://creativecommons.org/licenses/by/4.0/))

In vivo studies in rodents have reported liver fibrosis, alterations in the gut microbiome, testicular damage, and neurodevelopmental effects following chronic ingestion of microplastics [26]. In vitro experiments further reveal that nanoplastics can impair cellular metabolism, disrupt mitochondrial respiration, and trigger apoptosis in lung and gut epithelial cells [27]. The cellular uptake of nanoplastics via endocytosis and their interaction with organelles underscore their potential to bypass normal detoxification pathways.

An additional layer of complexity arises from the formation of the "eco-corona," a dynamic layer of biomolecules, proteins, and environmental pollutants that coat microplastics. This corona alters particle behavior and toxicity, making it difficult to predict outcomes solely on the basis of polymer type or size [28]. Furthermore, the potential for microplastics to act as carriers of antibiotic-resistant genes (ARGs) and pathogenic microorganisms is a growing concern in both environmental and clinical contexts [29].

Despite increasing evidence, human data remain sparse. Most findings stem from in vitro or animal studies, which may not fully translate to human health outcomes. Still, the convergence of toxicological, epidemiological, and environmental data indicates that microplastics are credible contributors to chronic disease burden. Risk assessments must now evolve to consider not only particle toxicity but also additive-driven and microbiome-mediated pathways.

6.3 Focus on Maternal and Child Health, Emerging Epidemiological Evidence

Microplastics are no longer an abstract environmental threat. Their presence in the most intimate human biological systems—such as the placenta, fetal tissues, and breast milk—has prompted an urgent call for interdisciplinary inquiry, particularly in maternal and child health. This section outlines the latest evidence on microplastic exposure during pregnancy, infancy, and early childhood, focusing on pathways, physiological responses, and early epidemiological insights.

6.3.1 Transplacental Exposure and Fetal Vulnerability

Emerging research confirms that microplastics can cross the placental barrier. Studies have identified polystyrene (PS), polyethylene (PE), and polyvinyl chloride (PVC) fragments in placental tissue, suggesting direct maternal-fetal transfer [30]. These particles are often small enough to bypass cellular defenses, entering maternal circulation through ingestion or inhalation before reaching the placenta. Their surface chemistry enables interactions with trophoblastic cells, potentially disrupting placental function and fetal nutrient exchange [31]. Animal models further reveal that prenatal exposure to polystyrene microplastics induces oxidative stress, inflammation, and apoptosis in fetal tissues, including the brain and liver [32, 33]. Additionally, microplastics can act as carriers of endocrine-disrupting chemicals like bisphenol A (BPA) and phthalates, further complicating fetal risk. These additives may alter fetal epigenetic programming, increasing the likelihood of metabolic, reproductive, and neurodevelopmental disorders in later life [34].

6.3.2 Breastfeeding and Neonatal Ingestion

Postnatal exposure begins immediately. Microplastics have been detected in human breast milk, raising concern about neonatal exposure during breastfeeding [35]. The detection of PE, PP, and PVC suggests a household origin, likely from food packaging, feeding equipment, or airborne contamination. Although concentrations appear low, the high sensitivity of the developing gut, immune system, and microbiome during infancy magnifies the potential consequences [36].

Moreover, bottle-fed infants may face greater risks. Li et al. [37] demonstrated that formula preparation in polypropylene bottles can release millions of microplastic particles per liter, depending on sterilization and shaking conditions. Given that neonatal organs are immature and highly permeable, even low-level exposure could influence immune development and trigger early inflammatory responses.

6.3.3 Microbiome Disruption and Immune Programming

The human gut microbiota is central to immune development, especially during the first 1000 days of life. Microplastics, acting as foreign bodies, can alter the composition and function of gut microbiota. In murine studies, polystyrene particles reduced the abundance of Bifidobacteria and Lactobacilli, while promoting pro-inflammatory taxa [38]. These shifts have been associated with increased gut permeability and systemic inflammation. In neonates, such perturbations could impair oral tolerance mechanisms, increasing the risk of allergies, autoimmune conditions, and inflammatory bowel disease [39].

Microplastic exposure may also affect the gut-brain axis. Neurobehavioral studies in zebrafish and rodents suggest altered neurotransmitter expression, locomotor activity, and anxiety-like behavior following early-life exposure to microplastics [40]. While human data are lacking, these findings underscore the neurodevelopmental sensitivity to microplastic-induced dysbiosis of the microbiota and inflammation.

6.3.4 Emerging Epidemiological Studies

Human data remain scarce but are slowly emerging. In a pilot study in Indonesia, researchers found microplastics in the placentas of 4 out of 6 women, with potential links to preterm birth and reduced fetal weight [19]. Another study by Leslie et al. [20] identified microplastic particles in human blood samples, including those from pregnant women. While these studies are not yet large enough to establish causality, they highlight the feasibility of exposure and biological uptake.

Epidemiological frameworks are beginning to integrate microplastics into exposome models, combining chemical and physical pollutant data with biomonitoring and birth outcomes [41]. Cross-sectional data suggest correlations between higher levels of microplastics in maternal urine and markers of oxidative stress and inflammation, although longitudinal data are needed.

6.3.5 Occupational Exposure Among Pregnant Women

Women working in industries with high levels of microplastic exposure—such as textiles, recycling, or food packaging—may experience occupational exposure levels several magnitudes higher than those in the general population. Inhalation of airborne microfibers has been documented in factory workers, and autopsy samples reveal accumulation in lung tissue [2]. Given the increased respiratory rate during pregnancy and the physiological changes that affect detoxification pathways, occupational exposure poses a disproportionate risk. Protective interventions and exposure guidelines remain absent in most regulatory settings.

6.3.6 Critical Gaps and Research Priorities

Despite growing evidence, several gaps limit the translation of research into public health policy. These include:

- **Standardization of detection methods:** Variability in sample collection, preparation, and identification techniques complicates cross-study comparisons.
- **Exposure quantification:** Reliable metrics for maternal and neonatal intake remain limited.
- **Longitudinal cohorts:** Most current studies are cross-sectional or animal-based. Prospective birth cohorts with repeated biological sampling are urgently needed.
- **Disentangling additive toxicity:** The health effects of microplastics must be separated from those of co-occurring plasticizers, pollutants, and microbes.

Developing sensitive, non-invasive tools for monitoring maternal and neonatal exposure to microplastics will be key to enabling evidence-based interventions and risk assessments.

6.4 Understanding Additive-Driven Toxicity Versus Particle Toxicity

Microplastics pose health risks not only because of their physical presence but also because of the diverse mix of chemical additives and environmental contaminants they can carry or adsorb. Recent toxicological research increasingly distinguishes two interconnected yet mechanistically distinct exposure pathways: particle toxicity, which originates from the intrinsic physical characteristics of microplastic particles, and additive-driven toxicity, which results from the release or transfer of associated chemicals into biological systems [42]. As illustrated in Fig. 6.3, particle toxicity is primarily governed by physical attributes, such as particle size, morphology, and surface reactivity. Fibers, fragments, and spherical particles can interact directly with biological membranes, causing physical irritation, membrane disruption, and cellular stress responses. These interactions may trigger the generation of

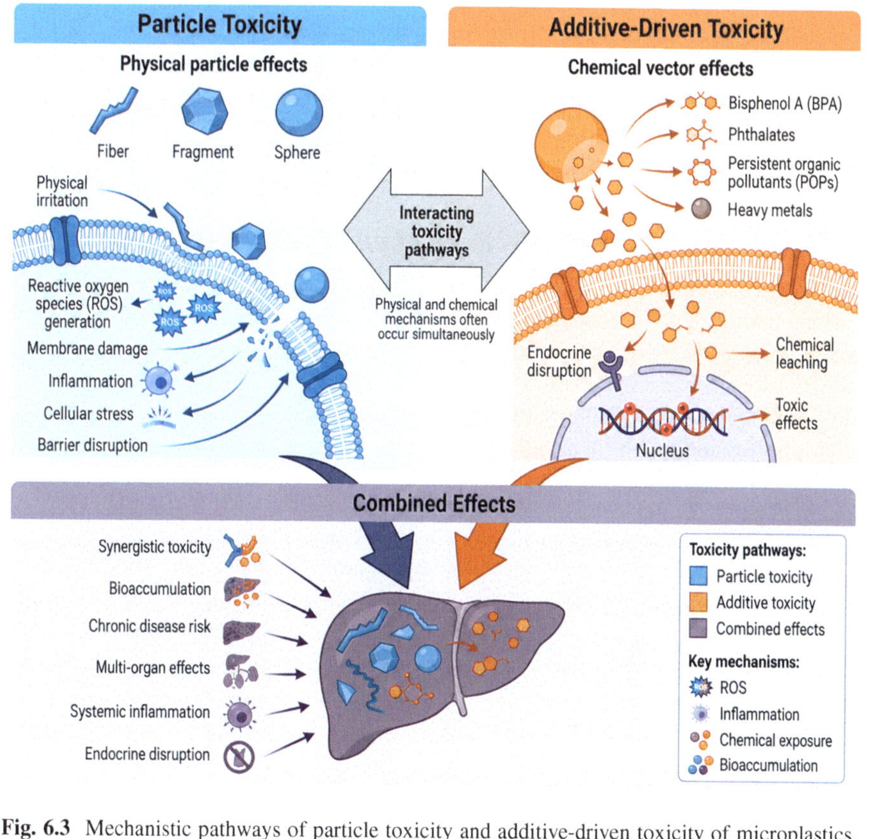

Fig. 6.3 Mechanistic pathways of particle toxicity and additive-driven toxicity of microplastics. (Conceptual framework developed by the authors and illustrated using digital design tools)

reactive oxygen species, leading to oxidative stress, inflammation, and disruption of epithelial barriers that normally protect tissues from external contaminants.

In contrast, additive-driven toxicity arises from the role of microplastics as chemical vectors [43]. The figure shows how plastics can release intrinsic additives, such as bisphenol A and phthalates, while also transporting sorbed environmental contaminants, including persistent organic pollutants and heavy metals. Once released into biological environments, these substances may interfere with endocrine signaling, disrupt gene regulation processes, and induce toxic cellular responses. Chemical leaching from ingested or inhaled particles, therefore, represents an important secondary mechanism of toxicity that operates alongside physical particle effects.

Importantly, the figure emphasizes that these pathways rarely occur in isolation. Instead, physical and chemical mechanisms frequently interact, producing combined biological effects that may exceed the sum of their individual impacts. These combined effects include synergistic toxicity, in which physical damage enhances chemical uptake and bioaccumulation following repeated exposure [44]. The diagram further highlights downstream systemic consequences, including chronic inflammation, multi-organ stress, endocrine disruption, and elevated long-term disease risk. By presenting these interacting pathways, the figure underscores the need to view microplastic toxicity as a multi-mechanism phenomenon in which particle effects and chemical exposures jointly determine biological outcomes, rather than as separate risk categories.

6.4.1 Particle Toxicity: Mechanical and Cellular Interface

Particle toxicity refers to the damage caused by the physical presence of microplastics in the body. Due to their small size, especially at the nanoscale, these particles can translocate across epithelial barriers and accumulate in tissues, such as the lungs, intestines, liver, kidneys, placenta, and brain [19, 20]. The shape and size of microplastics influence their interactions with cells, with fibrous and irregularly shaped particles triggering stronger inflammatory and immune responses than spherical particles [45].

Once internalized, microplastics can cause oxidative stress, damage to cellular membranes, disruption of tight junctions, and genotoxic effects [45]. Chronic exposure in murine models leads to intestinal barrier dysfunction and microbiome imbalance, highlighting subclinical damage that may accumulate over time [46]. Respiratory exposure to airborne microplastics has been linked to alveolar inflammation, fibrosis, and cytotoxicity in lung epithelial cells, particularly with prolonged occupational exposure [47]. These effects occur even when plastics are inert, indicating that particle-cell interactions are inherently toxic.

6.4.2 Additive-Driven Toxicity: A Trojan Horse Mechanism

Unlike inert particles, most consumer plastics contain a range of additives—plasticizers, flame retardants, antioxidants, UV stabilizers, colorants, and antimicrobials—used to enhance product performance. These additives are not covalently bound to the polymer matrix, making them prone to leaching into surrounding tissues once internalized [48]. Phthalates, bisphenol A (BPA), polybrominated diphenyl ethers (PBDEs), and organotins are among the most studied compounds, known for their endocrine-disrupting and neurotoxic properties [49].

Moreover, environmental microplastics act as vectors for adsorbed pollutants, including heavy metals (e.g., lead and cadmium), polychlorinated biphenyls (PCBs), polycyclic aromatic hydrocarbons (PAHs), and antibiotic-resistant genes (ARGs) [50]. This phenomenon creates a Trojan Horse effect, where the microplastic delivers harmful chemicals deep into the body's systems. Studies using gut and liver organoids have shown that additive-laden microplastics disrupt lipid metabolism, induce mitochondrial dysfunction, and provoke cytokine storms [36].

6.4.3 Interplay Between Particle and Chemical Effects

Disentangling particle and additive toxicity is challenging, as these pathways often operate in tandem. Recent in vitro studies demonstrated that pristine polystyrene beads were less toxic than those aged in seawater or contaminated with PAHs, which caused significant DNA fragmentation and apoptosis in human cell lines [51]. This implies that environmental weathering and chemical sorption modulate toxicity profiles.

Furthermore, particle toxicity may facilitate chemical toxicity by increasing epithelial permeability, enabling greater absorption of additives and co-contaminants [52]. Conversely, some additives, such as antimicrobial agents, may alter gut microbial communities, thereby exacerbating inflammatory responses triggered by the particle core [53].

6.4.4 Dose, Size, and Polymer Type Matter

Toxicity depends on multiple variables: particle size, polymer type, exposure route, and chemical load. Nanoscale particles have greater surface area-to-volume ratios, facilitating both tissue penetration and chemical release. Polyvinyl chloride (PVC) and polyethylene terephthalate (PET), for instance, tend to adsorb more heavy metals and phthalates than polypropylene (PP), due to their surface characteristics [54].

Inhalation routes tend to deliver microplastics to the lung-brain axis, while ingestion leads to gastrointestinal, hepatic, and systemic exposure. Occupational settings, such as textile factories and plastic recycling facilities, present high-exposure environments with frequent co-occurrence of microplastics and chemical fumes, amplifying both particle and additive toxicity risks [2].

6.4.5 Challenges in Risk Assessment and Regulation

Traditional toxicology evaluates chemicals and particles separately, yet microplastics require a hybrid assessment model. Regulatory agencies face a dilemma in defining acceptable exposure thresholds, as microplastics span a wide range of particle sizes and chemical profiles and exhibit variable bioavailability. Current assays may underestimate synergistic effects between particle and chemical toxicity, especially under chronic low-dose exposure relevant to real-world scenarios [55].

There is a pressing need for standardized testing protocols that simulate environmental aging, ingestion, and inhalation patterns. Animal models and human cell lines should incorporate both pristine and environmentally weathered microplastics to reflect real-world exposures. Moreover, additive migration studies must account for the cumulative effects of mixture toxicity, as humans rarely encounter a single contaminant in isolation.

6.5 Chapter Synthesis

This chapter establishes that microplastic pollution represents an emerging human health concern that extends beyond environmental contamination. Continuous exposure occurs through ingestion, inhalation, and dermal contact, placing human populations in direct contact with these particles throughout daily life. Evidence from food systems, drinking water, atmospheric deposition, and occupational settings shows that exposure occurs across both developed and developing regions.

Current toxicological evidence indicates that health risks arise through multiple mechanisms. Physical particle effects, chemical toxicity from plastic additives, and the transport of environmental contaminants all contribute to potential biological responses. Laboratory studies report inflammation, oxidative stress, cellular damage, and disruptions in metabolic processes following exposure to microplastics and associated chemicals. While long-term epidemiological evidence continues to develop, existing toxicological data justify precautionary concern and further investigation.

Particular attention is required for vulnerable populations. Maternal and child health research suggests that early life exposure may influence developmental processes, immune responses, and long-term health outcomes. Evidence of microplastic presence in human tissues and biological fluids underscores the importance of understanding exposure timing, accumulation pathways, and potential life-stage sensitivities. These findings reinforce the need for interdisciplinary collaboration among environmental scientists, toxicologists, epidemiologists, and public health specialists.

The distinction between particle-driven toxicity and additive-driven toxicity further clarifies risk pathways. Microplastics function as both physical stressors and chemical vectors. Plasticizers, stabilizers, flame retardants, and sorbed pollutants introduce additional toxicological complexity. Understanding the relative contributions of particle effects and chemical effects remains essential for accurate risk assessment and regulatory response.

This chapter, therefore, positions microplastic pollution within the broader context of environmental health science. Strengthening exposure assessment, advancing toxicological research, and improving biomonitoring approaches remain necessary to clarify risks and support science-based public health strategies. Continued research will play an important role in guiding preventive measures and informing global health discussions related to plastic pollution.

Human health impacts also reflect broader societal conditions. Exposure patterns often mirror inequalities in infrastructure, waste management capacity, occupational safety, and access to environmental protection. These realities connect microplastic pollution to questions of economic burden, social vulnerability, and environmental justice.

The next chapter expands the discussion toward the economic and social dimensions of microplastic pollution. Attention shifts toward the financial costs of environmental contamination, the distribution of risks across communities, and the structural factors that influence vulnerability, particularly in rapidly developing regions. This perspective situates microplastic pollution within the wider context of sustainable development and social resilience.

References

1. Roman, J., Gondko, D., Dębiec, P., & Pietrzak, N. (2024). The hidden health crisis: Microplastics and their medical consequences. *Journal of Education Health and Sport, 72*, 51690. https://doi.org/10.12775/jehs.2024.72.51690
2. Prata, J. C., da Costa, J. P., Lopes, I., Duarte, A. C., & Rocha-Santos, T. (2020). Environmental exposure to microplastics: An overview on possible human health effects. *Science of the Total Environment, 702*, 134455. https://doi.org/10.1016/j.scitotenv.2019.134455
3. Sharma, S., & Chatterjee, S. (2017). Microplastic pollution, a threat to marine ecosystem and human health: A short review. *Environmental Science and Pollution Research, 24*, 21530–21547. https://doi.org/10.1007/s11356-017-9910-8
4. Bhardwaj, G., Abdulkadhim, M., Joshi, K., Wankhede, L., Das, R. K., & Brar, S. K. (2025). Exposure pathways, systemic distribution, and health implications of micro- and nanoplastics in humans. *Applied Sciences, 15*(16), 8813. https://doi.org/10.3390/app15168813
5. Karami, A., Golieskardi, A., Choo, C. K., Larat, V., Karbalaei, S., & Salamatinia, B. (2018). Microplastic and mesoplastic contamination in canned sardines and sprats. *Science of the Total Environment, 612*, 1380–1386. https://doi.org/10.1016/j.scitotenv.2017.09.005
6. Cox, K. D., Covernton, G. A., Davies, H. L., Dower, J. F., Juanes, F., & Dudas, S. E. (2019). Human consumption of microplastics. *Environmental Science & Technology, 53*(12), 7068–7074. https://doi.org/10.1021/acs.est.9b01517

7. Barboza, L. G. A., Dick Vethaak, A., Lavorante, B. R. B. O., Lundebye, A.-K., & Guilhermino, L. (2018). Marine microplastic debris: An emerging issue for food security, food safety and human health. *Marine Pollution Bulletin, 133*, 336–348. https://doi.org/10.1016/j.marpolbul.2018.05.047
8. Li, J., Liu, H., & Chen, J. P. (2018). Microplastics in freshwater systems: A review on occurrence, environmental effects, and methods for microplastics detection. *Water Research, 137*, 362–374. https://doi.org/10.1016/j.watres.2017.12.056
9. Mason, S. A., Welch, V., & Neratko, J. (2018). Synthetic polymer contamination in bottled water. *Frontiers in Chemistry, 6*, 407. https://doi.org/10.3389/fchem.2018.00407
10. Schymanski, D., Goldbeck, C., Humpf, H. U., & Fürst, P. (2018). Analysis of microplastics in water by micro-Raman spectroscopy: Release of plastic particles from different packaging into mineral water. *Water Research, 129*, 154–162. https://doi.org/10.1016/j.watres.2017.11.011
11. Fadare, O. O., Wan, B., Guo, L.-H., & Zhao, L. (2020). Microplastics from consumer plastic food containers: Are we consuming it? *Chemosphere, 253*, 126787. https://doi.org/10.1016/j.chemosphere.2020.126787
12. Dris, R., Gasperi, J., Rocher, V., Saad, M., Renault, N., & Tassin, B. (2015). Microplastic contamination in an urban area: A case study in Greater Paris. *Environmental Chemistry, 12*(5), 592–599. https://doi.org/10.1071/EN14167
13. Liu, K., Wang, X., Fang, T., Xu, P., Zhu, L., & Li, D. (2019). Source and potential risk assessment of suspended atmospheric microplastics in Shanghai. *Science of the Total Environment, 675*, 462–471. https://doi.org/10.1016/j.scitotenv.2019.04.110
14. Vianello, A., Jensen, R. L., Liu, L., & Vollertsen, J. (2019). Simulating human exposure to indoor airborne microplastics using a breathing thermal manikin. *Scientific Reports, 9*, 8670. https://doi.org/10.1038/s41598-019-45054-w
15. Wright, S. L., & Kelly, F. J. (2017). Plastic and human health: A micro issue? *Environmental Science & Technology, 51*(12), 6634–6647. https://doi.org/10.1021/acs.est.7b00423
16. Prata, J. C. (2018). Airborne microplastics: Consequences to human health? *Environmental Pollution, 234*, 115–126. https://doi.org/10.1016/j.envpol.2017.11.043
17. Gasperi, J., Wright, S. L., Dris, R., Collard, F., Mandin, C., Guerrouache, M., Langlois, V., Kelly, F. J., & Tassin, B. (2018). Microplastics in air: Are we breathing it in? *Current Opinion in Environmental Science & Health, 1*, 1–5. https://doi.org/10.1016/j.coesh.2017.10.002
18. Campanale, C., Massarelli, C., Savino, I., Locaputo, V., & Uricchio, V. F. (2020). A detailed review study on potential effects of microplastics and additives of concern on human health. *International Journal of Environmental Research and Public Health, 17*(4), 1212. https://doi.org/10.3390/ijerph17041212
19. Ragusa, A., Svelato, A., Santacroce, C., Catalano, P., Notarstefano, V., Carnevali, O., Papa, F., Rongioletti, M. C. A., Baiocco, F., Draghi, S., D'Amore, E., Rinaldo, D., Matta, M., & Giorgini, E. (2021). Plasticenta: First evidence of microplastics in human placenta. *Environment International, 146*, 106274. https://doi.org/10.1016/j.envint.2020.106274
20. Leslie, H. A., van Velzen, M. J. M., Brandsma, S. H., Vethaak, A. D., Garcia-Vallejo, J. J., & Lamoree, M. H. (2022). Discovery and quantification of plastic particle pollution in human blood. *Environment International, 163*, 107199. https://doi.org/10.1016/j.envint.2022.107199
21. Yee, M. S.-L., Hui, L.-W., Looi, C. K., Lim, W.-M., Wong, S.-F., Kok, Y.-Y., Tan, B.-K., Wong, C.-Y., & Leong, C.-O. (2021). Impact of microplastics and nanoplastics on human health. *Nanomaterials, 11*(2), 496. https://doi.org/10.3390/nano11020496
22. Danopoulos, E., Twiddy, M., West, R. M., & Rotchell, J. M. (2021). A rapid review and meta-regression analyses of the toxicological impacts of microplastic exposure in human cells. *Journal of Hazardous Materials, 427*, 127861. https://doi.org/10.1016/j.jhazmat.2021.127861
23. Galloway, T. S. (2015). Micro- and nano-plastics and human health. In M. Bergmann, L. Gutow, & M. Klages (Eds.), *Marine anthropogenic litter* (pp. 343–366). Springer. https://doi.org/10.1007/978-3-319-16510-3_13

24. Hahladakis, J. N., Velis, C. A., Weber, R., Iacovidou, E., & Purnell, P. (2018). An overview of chemical additives present in plastics: Migration, release, fate and environmental impact during their use, disposal and recycling. *Journal of Hazardous Materials, 344*, 179–199. https://doi.org/10.1016/j.jhazmat.2017.10.014
25. Gao, H., Chen, D., & Zang, M. (2021). Association between phthalate exposure and insulin resistance: A systematic review and meta-analysis update. *Environmental Science and Pollution Research, 28*(40), 55967–55980. https://doi.org/10.1007/s11356-021-16252-9
26. Jin, Y., Xia, J., Pan, Z., Yang, J., Wang, W., & Fu, Z. (2018). Polystyrene microplastics induce microbiota dysbiosis and inflammation in the gut of adult zebrafish. *Environmental Pollution, 235*, 322–329. https://doi.org/10.1016/j.envpol.2017.12.088
27. Schirinzi, G. F., Pérez-Pomeda, I., Sanchís, J., Rossini, C., Farré, M., & Barceló, D. (2017). Cytotoxic effects of commonly used nanomaterials and microplastics on cerebral and epithelial human cells. *Environmental Research, 159*, 579–587. https://doi.org/10.1016/j.envres.2017.08.043
28. Cao, J., Yang, Q., Jiang, J., Dalu, T., Kadushkin, A., Singh, J., Fakhrullin, R., Wang, F., Cai, X., & Li, R. (2022). Coronas of micro/nano plastics: A key determinant in their risk assessments. *Particle and Fibre Toxicology, 19*(1), 55. https://doi.org/10.1186/s12989-022-00492-9
29. Li, J., Zhang, K., & Zhang, H. (2018). Adsorption of antibiotics on microplastics. *Environmental Pollution, 237*, 460–467. https://doi.org/10.1016/j.envpol.2018.02.050
30. Medley, E. A., Spratlen, M. J., Yan, B., Herbstman, J. B., & Deyssenroth, M. A. (2023). A systematic review of the placental translocation of micro- and nanoplastics. *Current Environmental Health Reports, 10*(2), 99. https://doi.org/10.1007/s40572-023-00391-x
31. Braun, T., Ehrlich, L., Henrich, W., Koeppel, S., Lomako, I., Schwabl, P., & Liebmann, B. (2021). Detection of microplastic in human placenta and meconium in a clinical setting. *Pharmaceutics, 13*(7), 921. https://doi.org/10.3390/pharmaceutics13070921
32. Jin, Y., Lu, L., Tu, W., Luo, T., & Fu, Z. (2019). Impacts of polystyrene microplastic on the gut barrier, microbiota and metabolism of mice. *Science of the Total Environment, 649*, 308–317. https://doi.org/10.1016/j.scitotenv.2018.08.353
33. Aghaei, Z., Sled, J. G., Kingdom, J., Baschat, A., Helm, P. A., Jobst, K. J., & Cahill, L. S. (2022). Maternal exposure to polystyrene micro- and nanoplastics causes fetal growth restriction in mice. *Environmental Science & Technology Letters, 9*(5), 426. https://doi.org/10.1021/acs.estlett.2c00186
34. Barboza, L. G. A., Lopes, C., Oliveira, P., Bessa, F., Otero, V., Henriques, B., Raimundo, J., Caetano, M., Vale, C., & Guilhermino, L. (2020). Microplastics in wild fish from North East Atlantic Ocean and its potential for causing neurotoxic effects, lipid oxidative damage, and human health risks associated with ingestion exposure. *Science of the Total Environment, 717*, 134625. https://doi.org/10.1016/j.scitotenv.2019.134625
35. Ragusa, A., Notarstefano, V., Svelato, A., Belloni, A., Gioacchini, G., Blondeel, C., Zucchelli, E., Luca, C. D., D'Avino, S., Gulotta, A., Carnevali, O., & Giorgini, E. (2022). Raman microspectroscopy detection and characterisation of microplastics in human breastmilk. *Polymers, 14*(13), 2700. https://doi.org/10.3390/polym14132700
36. Fackelmann, G., & Sommer, S. (2019). Microplastics and the gut microbiome: How chronically exposed species may suffer from gut dysbiosis. *Marine Pollution Bulletin, 143*, 193–203. https://doi.org/10.1016/j.marpolbul.2019.04.030
37. Li, D., Shi, Y., Yang, L., Xiao, L., Kehoe, D. K., Gun'ko, Y. K., Boland, J. J., & Wang, J. J. (2020). Microplastic release from the degradation of polypropylene feeding bottles during infant formula preparation. *Nature Food, 1*(11), 746. https://doi.org/10.1038/s43016-020-00171-y
38. Zhang, Z., Xu, M., Wang, L., Gu, W., Li, X., Han, Z., Fu, X., Wang, X., Xu, L., & Su, Z. (2023). Continuous oral exposure to micro- and nanoplastics induced gut microbiota dysbiosis, intestinal barrier and immune dysfunction in adult mice. *Environment International, 182*, 108353. https://doi.org/10.1016/j.envint.2023.108353

39. Thin, Z. S., Chew, J., Ong, T. Y. Y., Raja Ali, R. A., & Gew, L. T. (2025). Impact of microplastics on the human gut microbiome: A systematic review of microbial composition, diversity, and metabolic disruptions. *BMC Gastroenterology, 25*, 583. https://doi.org/10.1186/s12876-025-04140-2
40. Bora, S. S., Gogoi, R., Sharma, M. R., Anshu, A., Borah, M., Deka, P. C., Bora, J. R., Naorem, R. S., Das, J., & Teli, A. B. (2024). Microplastics and human health: Unveiling the gut microbiome disruption and chronic disease risks. *Frontiers in Cellular and Infection Microbiology, 14*, 1492759. https://doi.org/10.3389/fcimb.2024.1492759
41. Zurub, R. E., Cariaco, Y., Wade, M. G., & Bainbridge, S. (2024). Microplastics exposure: Implications for human fertility, pregnancy and child health. *Frontiers in Endocrinology, 14*, 1330396. https://doi.org/10.3389/fendo.2023.1330396
42. Hong, A. R., & Kim, J. S. (2024). Biological hazards of micro- and nanoplastic with adsorbents and additives. *Frontiers in Public Health, 12*, 1458727. https://doi.org/10.3389/fpubh.2024.1458727
43. Gouin, T., & Whelan, M. J. (2024). Evaluating microplastic particles as vectors of exposure for plastic additive chemicals using a food web model. *Microplastics and Nanoplastics, 4*, 21. https://doi.org/10.1186/s43591-024-00099-1
44. Dehaut, A., & Duflos, G. (2023). Microplastics (Chapter 7). In V. Verrez-Bagnis (Ed.), *Current challenges for the aquatic products processing industry*. Wiley. https://doi.org/10.1002/9781394264728.ch7
45. Zheng, L., Ma, X., Jin, Z., & Rao, Z. (2026). Microplastic-induced multi-organ toxicity: Cellular mechanisms and critical roles of organ crosstalk. *Frontiers in Public Health, 14*, 1746924. https://doi.org/10.3389/fpubh.2026.1746924
46. Chen, X., Zhuang, J., Chen, Q., Xu, L., Yue, X., & Qiao, D. (2022). Polyvinyl chloride microplastics induced gut barrier dysfunction, microbiota dysbiosis and metabolism disorder in adult mice. *Ecotoxicology and Environmental Safety, 241*, 113809. https://doi.org/10.1016/j.ecoenv.2022.113809
47. Picone, D., Carista, A., Piano, G., Manna, O. M., Burgio, S., Bucchieri, F., D'Amico, G., Pitruzzella, A., & Fucarino, A. (2025). Airborne microplastics: A Trojan horse for respiratory dysfunction and multiorgan damage. *Journal of Biological Research - Bollettino Della Società Italiana Di Biologia Sperimentale, 98*. https://doi.org/10.4081/jbr.2025.13945
48. Iftikhar, A., Qaiser, Z., Sarfraz, W., Ejaz, U., Aqeel, M., Rizvi, Z. F., & Khalid, N. (2024). Understanding the leaching of plastic additives and subsequent risks to ecosystems. *Water Emerging Contaminants & Nanoplastics, 3*(1), 5. https://doi.org/10.20517/wecn.2023.58
49. Zhang, B., Wu, D., Yang, X., Teng, J., Liu, Y., Zhang, C., Zhao, J., Yin, X., You, L., Liu, Y., & Wang, Q. (2019). Microplastic pollution in the surface sediments collected from Sishili Bay, North Yellow Sea, China. *Marine Pollution Bulletin, 141*, 9–15. https://doi.org/10.1016/j.marpolbul.2019.02.021
50. Rafa, N., Ahmed, B., Zohora, F. T., Bakya, J., Ahmed, S., Ahmed, S. F., Mofijur, M., Chowdhury, A. A., & Almomani, F. (2023). Microplastics as carriers of toxic pollutants: Source, transport, and toxicological effects. *Environmental Pollution, 343*, 123190. https://doi.org/10.1016/j.envpol.2023.123190
51. Almeida, M., Martins, M. A., Soares, A. M. V. M., Cuesta, A., & Oliveira, M. (2019). Polystyrene nanoplastics alter the cytotoxicity of human pharmaceuticals on marine fish cell lines. *Environmental Toxicology and Pharmacology, 69*, 57. https://doi.org/10.1016/j.etap.2019.03.019
52. Llorca, M., & Farré, M. (2021). Current insights into potential effects of micronanoplastics on human health by in-vitro tests. *Frontiers in Toxicology, 3, 752140*. https://doi.org/10.3389/ftox.2021.752140
53. Ghebretatios, M., Schaly, S., & Prakash, S. (2021). Nanoparticles in the food industry and their impact on human gut microbiome and diseases. *International Journal of Molecular Sciences, 22*(4), 1942. https://doi.org/10.3390/ijms22041942

54. Paul, M. B., Stock, V., Cara-Carmona, J., Lisicki, E., Shopova, S., Fessard, V., Braeuning, A., Sieg, H., & Böhmert, L. (2020). Micro- and nanoplastics—Current state of knowledge with the focus on oral uptake and toxicity. *Nanoscale Advances, 2*(10), 4350. https://doi.org/10.1039/d0na00539h
55. Liu, M., Liu, J., Xiong, F., Xu, K., Pu, Y., Huang, J., Zhang, J., Pu, Y., Sun, R., & Cheng, K. (2023). Research advances of microplastics and potential health risks of microplastics on terrestrial higher mammals: A bibliometric analysis and literature review. *Environmental Geochemistry and Health, 45*(6), 2803–2838. https://doi.org/10.1007/s10653-022-01458-8

Chapter 7
Economic and Social Dimensions

Introduction

Microplastics have emerged not only as an ecological and health concern but as a socioeconomic burden with deep structural implications [1, 2]. Their persistence in the environment has translated into escalating cleanup costs, strained waste management systems, and externalities that disproportionately affect marginalized communities [3, 4]. Governments across the globe, especially in developing regions, confront rising expenditures on coastal remediation, monitoring, and infrastructure development to manage plastic leakage and its microfragmentation, often exceeding available public resources and creating trade-offs with essential services, such as healthcare and education [5–7].

Moreover, microplastics intersect with the global climate crisis through shared fossil fuel dependencies and co-emissions associated with plastic production, transport, and degradation [1, 2]. Their accumulation in marine and terrestrial food systems influences labor, livelihoods, and community well-being, particularly among populations dependent on fisheries and agriculture [8, 9]. Social inequalities further compound these challenges, as low-income populations experience higher exposure due to inadequate waste infrastructure, informal settlements, and limited regulatory protection [3, 10].

In Southeast Asia, where plastic leakage into rivers and coastal waters is among the highest globally, weak waste governance and rapid urbanization exacerbate socioeconomic vulnerabilities [11, 12]. Understanding these interconnected economic and social dynamics is essential for designing globally coordinated, equity-centered responses that address the full societal burden of microplastics and enable effective, just, and sustainable interventions [2].

R. Y. Capangpangan, A. C. Alguno, *Microplastics in Focus*,
https://doi.org/10.1007/978-3-032-28978-0_7

7.1 The Cost of Cleanup: Financial Implications for Governments and Communities

Microplastics have become a pervasive and enduring environmental pollutant, posing significant economic burdens on national and local governments. Cleanup costs associated with plastic and microplastic pollution are growing exponentially as these pollutants infiltrate terrestrial and marine ecosystems, often beyond the reach of conventional waste management systems. The financial consequences of microplastic contamination span sectors, from direct removal and infrastructure investment costs to indirect costs affecting public health, fisheries, tourism, and urban resilience.

7.1.1 Direct Cleanup and Waste Management Costs

One of the most immediate financial impacts of microplastic pollution is the cost of environmental cleanup. In the United States alone, municipal governments collectively spend over $11.5 billion annually to manage litter, with plastic waste accounting for a major share [13]. In the European Union, coastal municipalities incur an estimated €630 million per year in costs for beach cleaning and waste disposal related to marine litter, of which plastics account for over 80% [14].

Microplastics complicate these efforts by requiring specialized removal methods. Traditional solid waste collection systems are ineffective for particles smaller than 5 mm, necessitating the deployment of advanced filtration in wastewater treatment facilities and sediment dredging in aquatic systems. According to Mitrano and Wohlleben (2020) [15], upgrading wastewater treatment plants to effectively capture microplastics may require significant capital expenditures, with estimated costs ranging from $5 million to $25 million per facility, depending on capacity and technology.

7.1.2 Impacts on Municipal Budgets and Public Infrastructure

For lower-income nations and municipalities, the financial burden is especially pronounced. Many coastal regions in Southeast Asia, including the Philippines, Vietnam, and Indonesia, lack adequate solid waste and wastewater infrastructure, resulting in high rates of plastic leakage into the ocean. These countries allocate a disproportionately high share of local budgets to basic waste collection while struggling to manage marine litter and stormwater runoff effectively [16].

Cleanup campaigns divert essential funds from health, education, and urban development. In Indonesia, the government allocated $1 billion over 4 years to address marine plastic pollution under the National Action Plan for Marine Debris

[17]. While this signals strong political commitment, it simultaneously highlights the scale of required investments—resources that could otherwise be directed to pressing socioeconomic needs.

Moreover, urban drainage systems clogged with microplastic-laden debris contribute to increased flooding and infrastructure damage, requiring further expenditure. For example, studies have linked plastic waste accumulation in storm drains in Manila to heightened flood risks during monsoons, amplifying maintenance and emergency response costs [2].

7.1.3 Losses to Coastal Tourism and Fisheries

Microplastics also impose indirect costs by degrading natural assets critical to local economies. Coastal and marine tourism sectors suffer when beaches and waterways are visibly polluted. In South Korea, surveys found that the presence of plastic debris on beaches could reduce tourist visitation by over 40%, leading to annual economic losses of $29 million in just one region [18].

In the fisheries sector, economic risks arise from reduced fish stock quality and productivity. Microplastics ingested by commercial fish and shellfish compromise food safety, marketability, and export potential. A UNEP report (2021) [19] estimated that marine plastic pollution costs the global fisheries and aquaculture industry over $1.3 billion annually in lost revenue and damaged gear [19].

In Southeast Asia, where fisheries are a primary livelihood for millions, microplastics threaten both subsistence and commercial activities. Contaminated seafood risks losing access to premium export markets with stricter food safety regulations, as exemplified by the European Union's growing scrutiny of microplastic levels in imported fish products [20].

7.1.4 Environmental Monitoring and Research Costs

Accurately tracking and assessing the extent of microplastic contamination is resource-intensive. Governments, academic institutions, and environmental agencies invest heavily in field sampling, laboratory analysis, and monitoring networks. Developing countries face unique constraints due to limited scientific capacity and access to standardized equipment. The absence of global harmonization in microplastic measurement adds further complexity, increasing costs through redundant studies and data inconsistencies [21].

To support evidence-based policymaking, some governments are integrating microplastic monitoring into national environmental frameworks. For instance, the European Marine Strategy Framework Directive now includes microplastics as an indicator of Good Environmental Status, requiring member states to finance long-term observation programs [22].

7.1.5 Community and Citizen Costs

Beyond governments, communities shoulder considerable costs, both economic and social. Volunteer cleanup drives, while symbolically powerful, often require coordination, logistics, and protective equipment that community groups must procure or crowdsource. Informal waste pickers, a key part of the plastic collection system in Southeast Asia, operate without adequate health safeguards and face economic precarity as markets for recycled plastic fluctuate due to contamination with microplastic particles [23].

Moreover, the psychosocial burden of living in polluted environments affects community cohesion, mental health, and sense of place. These intangible impacts, though harder to monetize, contribute to the overall societal cost of microplastic contamination.

7.1.6 Global Funding Gaps and Policy Externalities

Despite growing awareness, global funding remains insufficient. The UNEP estimates that cleaning up existing marine plastic pollution could cost up to $100 billion annually, yet current global plastic treaty negotiations lack binding financing mechanisms [19]. Meanwhile, many plastic-producing corporations externalize cleanup responsibilities, placing the burden on taxpayers and local authorities. Extended Producer Responsibility (EPR) schemes have emerged to reallocate costs, but implementation remains inconsistent across regions.

In response, new economic instruments are being piloted. For example, the Plastic Waste Reduction Standard enables companies to purchase verified plastic credits, funding collection and recycling efforts in developing countries. While promising, critics argue that such market-based approaches risk commodifying pollution and delaying systemic reform [24].

To better illustrate how microplastic pollution translates into economic burdens across environmental systems and governance structures, Fig. 7.1 presents an integrated conceptual framework linking pollution sources, environmental pathways, and the resulting direct, indirect, and governance-related costs. The framework shows how microplastics originating from plastic leakage, waste mismanagement, industrial activities, and consumer waste move through marine, wastewater, soil, and atmospheric systems. These interconnected pathways generate direct economic costs, such as cleanup operations, environmental monitoring, and infrastructure upgrades, as well as indirect costs, including fisheries losses, tourism decline, health expenditures, and flood damage. The figure also highlights governance and policy expenditures associated with waste management programs, research investments, and regulatory compliance, emphasizing that the financial pressures of microplastic pollution extend beyond remediation to include ecosystem degradation, public health protection, and long-term institutional responses.

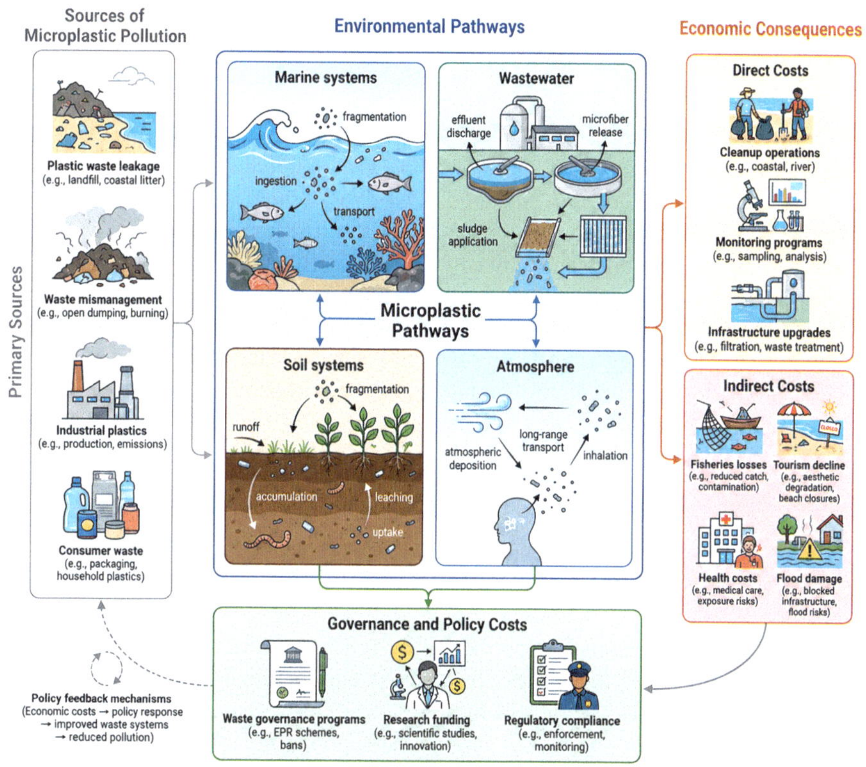

Fig. 7.1 Economic cost pathways of microplastic pollution. (Conceptual framework developed by the authors and illustrated using digital design tools)

7.1.7 *Toward Cost-Effective Solutions*

To reduce long-term financial burdens, investments must shift upstream toward plastic reduction, reuse, and innovation in materials. Implementing circular economy principles could decrease plastic production, mitigate leakage, and minimize cleanup needs. A study by the Pew Charitable Trusts and SYSTEMIQ (2020) [25] projects that adopting a comprehensive "reduce-reuse-recycle" approach could cut global plastic leakage by 80% and save governments $70 billion by 2040.

Strengthening regulatory enforcement, scaling decentralized waste treatment, and promoting product redesign can reduce the need for expensive reactive cleanups. Moreover, inclusive policymaking that integrates community voices, informal sectors, and marginalized groups will enhance both equity and effectiveness.

7.2 Microplastics and Climate Change: An Interconnected Crisis

The relationship between microplastics and climate change represents a deeply interwoven and underexplored nexus of global environmental concern. Both crises are symptoms of the world's overdependence on fossil fuels and unsustainable consumption patterns. Microplastics, formed from the fragmentation of plastic waste, embody the lifecycle emissions of plastic—from fossil fuel extraction to end-of-life decomposition—linking them intrinsically to the climate emergency [26]. At every stage of their life cycle, plastics contribute to greenhouse gas (GHG) emissions, with microplastics acting as both physical pollutants and vectors of broader carbon-related impacts. These multidirectional interactions illustrate a complex climate-plastic nexus that requires a systems perspective to fully understand the feedback mechanisms involved.

7.2.1 Shared Roots: Fossil Fuel Dependency

Plastic production accounts for approximately 4–8% of global oil consumption, and this figure is projected to increase by 2050 if current trends continue [27]. The extraction, transport, and refining of fossil fuels for plastic manufacturing emit large quantities of carbon dioxide (CO_2) and methane, potent greenhouse gases. Even in the post-consumer phase, plastic degradation in landfills and oceans can release methane and ethylene, contributing to climate forcing [28].

This fossil fuel dependency is especially critical in the context of climate negotiations. As some sectors decarbonize, the petrochemical industry is increasingly leaning on plastic as a growth area, undermining climate targets [29]. Thus, microplastics are not merely byproducts of pollution but indicators of a larger systemic failure in carbon governance.

7.2.2 Climate Feedback Loops and Oceanic Disruption

Microplastics affect critical carbon sinks such as oceans and soils, altering their biogeochemical functionality. The presence of microplastics in marine environments disrupts the biological pump—a process by which phytoplankton sequester atmospheric carbon and transport it to the deep ocean [30]. Laboratory studies show that microplastic ingestion affects zooplankton buoyancy, reproduction, and fecal pellet density, potentially impairing the sinking of carbon-rich organic matter [31]. This interference reduces oceanic carbon sequestration, exacerbating atmospheric CO_2 accumulation.

In cryospheric systems, microplastics deposited on snow and ice reduce surface albedo, accelerating melting and contributing to positive climate feedback loops [32]. These physical interactions amplify climate sensitivity in polar regions, where even marginal warming can lead to cascading effects on sea level rise and ecosystem collapse.

7.2.3 Emissions from Waste Management and Open Burning

Waste mismanagement, particularly in low- and middle-income countries, presents another intersection of microplastics and climate change. Inadequate waste infrastructure leads to open burning of plastics, a practice that emits CO_2, methane, black carbon, and other short-lived climate pollutants [33]. These emissions are not only hazardous to local air quality but also contribute significantly to global warming.

Plastic incineration—even under controlled conditions—has a substantial carbon footprint. According to Hopewell et al. (2009) [34], one metric ton of plastic incinerated can emit over two metric tons of CO_2 equivalent. The presence of microplastics in combusted waste streams complicates air quality control and poses occupational hazards for communities living near incinerators.

7.2.4 Interactions with Soil and Agricultural Systems

Microplastics in terrestrial ecosystems, particularly in agricultural lands, also affect climate change. Their presence affects soil structure, water retention, microbial diversity, and nutrient cycling [35]. Disruption of microbial communities can inhibit carbon sequestration in soils, further weakening a crucial terrestrial carbon sink. Moreover, synthetic fabrics and microplastic-laden biosolids applied as fertilizer may introduce persistent pollutants into soils, influencing nitrous oxide emissions and reducing resilience to drought and erosion.

This terrestrial-microplastic interaction is especially concerning in developing countries, where the application of untreated sewage sludge or wastewater for irrigation is common [36]. Without regulatory safeguards, microplastic accumulation in farmlands will continue to undermine both food security and carbon neutrality efforts.

7.2.5 Contribution to Atmospheric Pollution

Recent studies have shown that microplastics are now airborne, traveling vast distances and contributing to atmospheric particulates. These airborne microplastics may play a role in cloud condensation and radiative forcing, although their exact influence on climate dynamics is still being investigated [37]. In urban areas, tire wear particles and synthetic microfibers from textiles contribute to fine particulate matter (PM2.5), intensifying both public health and climate concerns.

Atmospheric transport pathways also link remote ecosystems to urban sources. For instance, microplastics have been detected in alpine and Arctic snow, suggesting long-range atmospheric movement, much like black carbon or other pollutants with climate relevance [38].

7.2.6 Economic and Policy Ramifications

The intertwined nature of microplastics and climate change creates dual policy burdens for governments, particularly in the Global South. Efforts to mitigate climate emissions are undermined by plastic proliferation and vice versa. Current climate financing mechanisms often overlook plastic waste management as a mitigation strategy, despite its clear relevance [39]. Integrating plastic lifecycle management into nationally determined contributions (NDCs) could offer co-benefits in emissions reductions and environmental protection. On the economic front, the failure to address this nexus incurs rising social costs. Coastal communities dependent on tourism and fisheries face ecosystem degradation and income losses, while governments bear mounting expenses in both climate adaptation and plastic remediation [40].

7.2.7 Toward Integrated Solutions

Solving the dual crisis requires cross-cutting strategies that view microplastic pollution as part of the broader planetary boundary framework. Circular economy approaches, particularly those that reduce virgin plastic production, can yield both emissions savings and pollution reduction. Enhanced monitoring, standardized metrics, and lifecycle assessment tools are essential for quantifying the true climate cost of plastics [5].

Ultimately, the link between microplastics and climate change is not merely a scientific abstraction but a lived reality with tangible environmental, economic, and societal consequences, particularly for vulnerable communities. As synthesized in Fig. 7.2, the interconnected relationships among plastic lifecycle emissions, climate processes, environmental disruptions, and policy responses demonstrate that microplastic pollution and climate change cannot be addressed in isolation. The framework shows how plastic production, transport, and waste burning contribute to greenhouse gas emissions and climate forcing, while persistent microplastics interact with environmental systems by disrupting ocean carbon sequestration processes, contributing to soil carbon loss, and undergoing atmospheric transport.

The figure further illustrates how climate processes, including carbon cycle disruption, albedo change, and radiative forcing, interact with microplastic dynamics

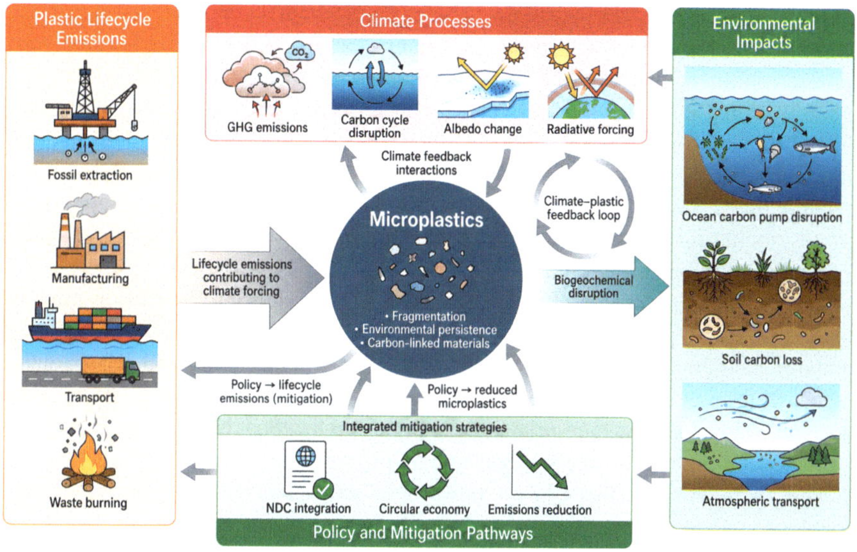

Fig. 7.2 Microplastics-climate change interaction framework. (Conceptual framework developed by the authors and illustrated using digital design tools)

through climate-plastic feedback loops that influence their distribution, persistence, and environmental impacts. These interactions highlight the need to view microplastics not only as pollutants but also as emerging components of global biogeochemical and climate systems.

Importantly, the framework also emphasizes that addressing these interconnected challenges requires systemic shifts in production, consumption, and governance anchored in sustainability, justice, and intergenerational equity. Integrated mitigation strategies, such as circular economy transitions, emissions-reduction policies, improved waste management systems, and the inclusion of plastics management within nationally determined contributions (NDCs), provide pathways to simultaneously reduce climate emissions and plastic pollution. By linking lifecycle emissions, climate interactions, environmental impacts, and policy solutions, the figure underscores the importance of coordinated, systems-based interventions that deliver both environmental protection and climate co-benefits.

7.3 Social Inequalities and Vulnerable Populations: Who Bears the Burden?

Microplastic pollution is not merely an environmental or technological problem; it is also a deeply social and distributive challenge. Vulnerable populations, particularly those in low-income, densely populated, and environmentally degraded regions, often bear a disproportionate share of the health, economic, and infrastructural burdens associated with plastic waste and microplastics. These burdens reinforce existing inequalities and are intensified by weak regulatory systems, poor urban planning, and chronic underinvestment in waste infrastructure. The unequal distribution of microplastic exposure reflects broader patterns of environmental injustice, where social vulnerability intersects with environmental risk. As illustrated in Fig. 7.3, microplastic exposure disproportionately affects vulnerable social groups, including low-income communities, women, children, waste workers, and Indigenous and coastal populations.

These groups experience heightened exposure through multiple environmental and occupational pathways, including inhalation of airborne particles, ingestion of contaminated food and water, contact with contaminated dust, and work-related exposure in waste-handling and recycling activities. Differences in occupation, geography, gender roles, age-related behaviors, and levels of institutional protection further shape these exposure patterns and amplify risk.

The framework further shows how these exposure pathways translate into broader social and health consequences, including increased health risks, livelihood vulnerability, and the amplification of existing inequalities. Importantly, the figure

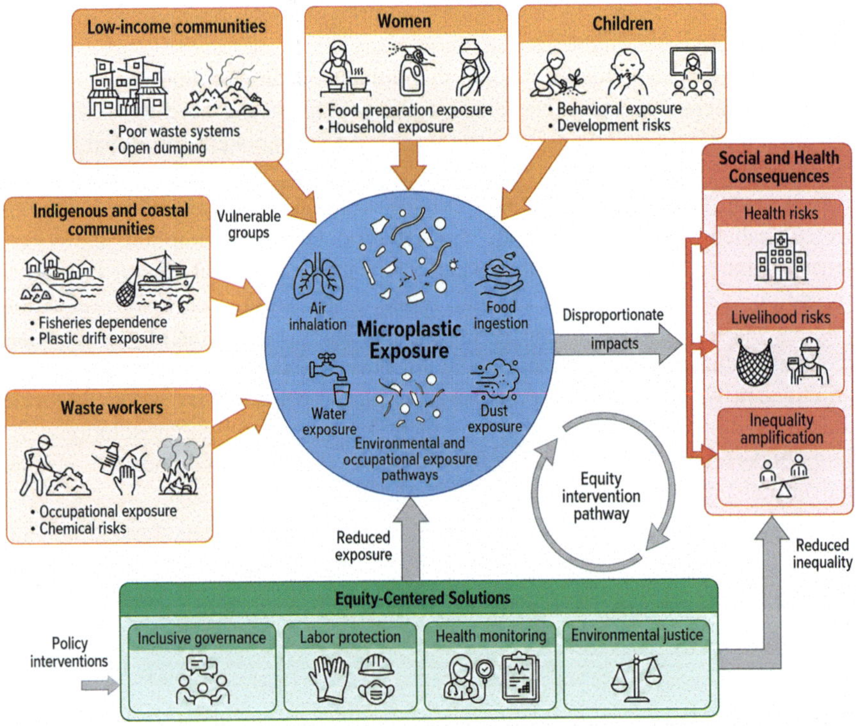

Fig. 7.3 Social distribution of microplastic exposure risks. (Conceptual framework developed by the authors and illustrated using digital design tools)

also highlights the role of equity-centered solutions in reducing these disparities. Policy interventions such as inclusive governance, labor protection, health monitoring, and environmental justice approaches provide pathways to reduce exposure, strengthen social protection, and promote more equitable environmental outcomes. By linking vulnerable groups, exposure pathways, societal consequences, and targeted policy responses, the figure emphasizes that effective microplastic governance must integrate scientific, social, and justice-oriented perspectives.

7.3.1 Disproportionate Exposure in Low-Income Communities

Low-income and marginalized populations are more likely to reside in areas with inadequate waste management systems, closer to open dumpsites, rivers with high plastic leakage, or industrial zones that contribute to plastic and microplastic emissions. In Southeast Asia, for example, communities in informal settlements along riverbanks or coastal areas often rely on polluted waterways for washing, cooking, and fishing—intensifying both ingestion and dermal exposure routes to microplastics and associated additives [41, 42]. Urban slums in Indonesia, the Philippines, and Vietnam have been shown to suffer from higher incidences of plastic pollution due to the absence of formal waste collection services, exposing children and adults to daily contact with plastic debris and microplastics embedded in soil, dust, and water systems.

Microplastic-contaminated dust and air particles have been identified in homes, schools, and markets in low-resource settings. Inhalation exposure is especially concerning in densely populated urban zones where waste is openly burned, releasing airborne particles and additives, such as phthalates and brominated flame retardants, many of which have been linked to respiratory and endocrine disorders [43, 44]. Indoor environments, including daycare centers and low-income housing units, may also serve as hotspots for airborne microplastics, particularly from synthetic textiles and degraded plastics.

7.3.2 Gendered and Age-Based Vulnerabilities

Women and children face distinct risks from microplastics. In many rural and peri-urban households, women are responsible for tasks, such as food preparation, water collection, and cleaning—roles that place them in closer contact with potentially contaminated media. Microplastics in drinking water and foodstuffs (especially fish, shellfish, and salt) pose ingestion risks, while skin contact during household chores may result in dermal exposure. Moreover, the maternal-fetal transfer of microplastics through the placenta, although still under study, has been documented in preliminary human samples, raising concerns for prenatal development [45, 46].

Children, too, are at heightened risk. Their physiological immaturity and behavioral tendencies—such as playing in contaminated soils, mouthing objects, and breathing closer to the ground—render them more susceptible to both oral and inhalation exposure pathways [47]. In schools with deteriorating infrastructure or open drainage, these risks are amplified, especially where plastic waste accumulates and breaks down in situ.

7.3.3 Occupational Exposure and the Informal Waste Sector

The informal waste economy—recycling, scavenging, and sorting—serves as a vital livelihood for millions, particularly in the Global South. However, this economic activity exposes workers to high concentrations of microplastics through constant contact with degraded plastic materials, contaminated dust, and fumes from burning plastic. These workers, often women and children, operate with little to no protective equipment and in unregulated conditions. A study in Ghana found that waste pickers at urban dumpsites exhibited elevated biomarkers for phthalates and bisphenol A, chemicals commonly associated with microplastics [48, 49].

In Southeast Asia, the informal waste economy plays an outsized role in plastic recovery. Yet, policies and extended producer responsibility (EPR) frameworks often fail to formally include these actors in decision-making or provide them with safety nets and health protection. As microplastics increasingly infiltrate recycling streams, particularly from multi-layered or degraded plastics, risks to these workers multiply due to the potential release of nanoplastics and hazardous additives during processing [50].

7.3.4 Environmental Racism and Locational Disparities

In many parts of the world, communities already facing environmental injustice—such as Indigenous populations, ethnic minorities, or migrants—are disproportionately located near landfills, ports, plastic processing hubs, or industrial parks that serve as microplastic emission hotspots. In the United States, for instance, African American and Hispanic communities are more likely to live near incineration facilities and industrial plastic plants, with documented health disparities linked to toxic emissions [51].

In Southeast Asia, remote island communities bear the brunt of plastic drift from global currents. These communities often lack the infrastructure to manage even local waste, let alone address transboundary plastic loads. As a result, microplastics accumulate in coastal sediments, reefs, and fisheries that sustain local livelihoods [52]. Here, environmental injustice takes on a global dimension—pollution generated by industrialized nations disproportionately affects distant, vulnerable communities with the least capacity for remediation or policy advocacy.

7.3.5 Lack of Policy Protections and Institutional Responses

Despite growing evidence of harm, most microplastic governance frameworks remain technocratic and ignore social justice dimensions. Risk assessments focus on chemical toxicology, with limited integration of sociological data, such as differential exposure, livelihood impacts, or cumulative vulnerabilities. Few countries explicitly incorporate equity-focused interventions in their plastic action plans or invest in community-based monitoring, capacity building, or healthcare infrastructure responsive to microplastic exposure.

Moreover, data disaggregation by class, gender, geography, or occupation remains scarce. This invisibility in research and policy further marginalizes vulnerable groups, delaying tailored interventions. Without systematic inclusion of social vulnerability metrics, mitigation strategies risk perpetuating existing inequalities—or even worsening them by concentrating cleanup or recycling costs on communities least responsible for the pollution.

7.3.6 Toward Equity-Centered Solutions

Addressing the uneven burden of microplastic pollution requires a shift toward equity-centered frameworks. This entails:

- *Participatory governance*, where frontline communities are involved in shaping policies and risk assessments.
- *Targeted funding* for waste infrastructure upgrades in underserved areas.
- *Labor protections* for informal waste workers and recyclers.
- *Health surveillance* systems that monitor microplastic exposure across demographic groups.
- *Environmental education* campaigns that account for local language, culture, and accessibility.

International mechanisms such as the Plastics Treaty must also embed principles of environmental justice and include binding provisions for protecting at-risk populations. Climate justice and plastic justice are inseparable—vulnerable populations cannot be afterthoughts in a just transition to a circular and toxin-free economy.

7.4 Southeast Asian Challenges in Waste Infrastructure and Management

Southeast Asia, one of the fastest-growing economic regions in the world, is also at the epicenter of the global plastic pollution crisis. As urban populations expand and consumption patterns shift, the region's waste management systems are facing unprecedented strain. Microplastics—a byproduct of this systemic challenge—are not only a symptom of poor waste governance but also a catalyst for environmental and social vulnerability. Addressing the challenges of waste infrastructure and microplastic leakage in Southeast Asia requires a systemic, inclusive, and context-aware approach that acknowledges the heterogeneity of governance systems, financial capacity, and ecological risk across countries.

7.4.1 Weak Waste Infrastructure and Mismanaged Plastics

Southeast Asia is home to several of the top plastic-polluting countries globally, including the Philippines, Indonesia, Vietnam, and Thailand. A major driver of this is the large volume of mismanaged plastic waste, defined as waste that is either inadequately disposed (e.g., open dumping and uncontrolled landfilling) or not collected at all. According to estimates, Indonesia alone generates approximately 6.8 million tons of plastic waste annually, with over 600,000 tons leaking into aquatic environments each year [53].

The root of this leakage lies in fragmented and under-resourced municipal waste systems. Most cities and rural areas in the region lack comprehensive waste collection coverage, sorting facilities, or final disposal infrastructure. In the Philippines, only 30% of municipalities operate sanitary landfills, and a significant portion of plastic waste is burned or dumped into waterways [54]. Even in urban centers like Jakarta and Manila, the rise in single-use plastics has far outpaced the capacity of existing waste systems to manage them effectively.

7.4.2 Informal Waste Sector and Structural Inequalities

The waste economy in many Southeast Asian countries is highly informal. Waste pickers play a crucial role in recovering recyclable materials, but their work often occurs under hazardous and insecure conditions. While they are instrumental in reducing plastic leakage into the environment, these workers are rarely integrated into formal waste governance frameworks, leading to systemic underinvestment in their safety and livelihoods [55].

The exclusion of the informal sector from municipal waste planning not only weakens overall system efficiency but also compounds social inequality. Women, children, and ethnic minorities are disproportionately represented among waste workers, and exposure to plastic residues, including microplastics and toxic additives, has been linked to respiratory problems and reproductive disorders [10, 50]. Without social protection or access to healthcare, these communities bear a disproportionate share of the health burden associated with poor plastic waste management.

7.4.3 *Limited Regional Coordination and Governance Gaps*

One of the critical barriers to effective microplastic mitigation in Southeast Asia is the fragmented governance landscape. Countries vary widely in the maturity of their regulatory frameworks. For example, while Malaysia has initiated the "Roadmap Towards Zero Single-Use Plastics," enforcement mechanisms remain weak. In contrast, the Philippines' Republic Act 9003 (Ecological Solid Waste Management Act) mandates waste segregation at source, but compliance remains patchy due to a lack of local enforcement capacity and sustained public awareness campaigns [56–58].

Regional coordination through ASEAN has been slow, though recent efforts such as the ASEAN Regional Action Plan for Combating Marine Debris in the ASEAN Member States (2021–2025) have created a blueprint for harmonized policy actions, including extended producer responsibility (EPR), sustainable consumption, and investment in waste infrastructure [53, 59]. However, the action plan remains underfunded and largely aspirational, with limited binding commitments from member states.

7.4.4 *Financial Constraints and Trade-Offs in Local Governments*

Local governments, which are primarily responsible for waste management in most Southeast Asian countries, face chronic underfunding. Waste management is often the single largest line-item expenditure in municipal budgets, accounting for 20–50% of funds while covering only 60% of urban waste streams [60].

The financial constraints are exacerbated by increasing waste volumes and rising operational costs for collection, transport, and landfill maintenance. Many municipalities lack the capital to invest in modern material recovery facilities (MRFs), composting units, or waste-to-energy technologies. Consequently, plastic and microplastic pollutants often bypass controlled disposal systems and accumulate in open dumpsites, drainage systems, and coastal zones. Trade-offs are common: governments are forced to choose between improving health services, education, and waste management.

Furthermore, international development financing for plastic waste management in Southeast Asia has historically been limited and fragmented. Most donor efforts are pilot-scale and fail to achieve lasting systemic transformation due to weak ownership and coordination.

7.4.5 *Climate and Disaster Amplifiers*

Southeast Asia is highly vulnerable to climate-related disasters, such as typhoons, floods, and landslides. These events exacerbate plastic pollution by overwhelming already fragile waste systems. During major floods, uncollected plastic waste is flushed into rivers and oceans, where it fragments into microplastics. For instance, Typhoon Haiyan (Yolanda) in 2013 resulted in the release of thousands of tons of plastic debris across the Philippine archipelago [61, 62].

Moreover, the open burning of plastics—a widespread practice in low-income and rural areas—releases black carbon, persistent organic pollutants, and greenhouse gases that contribute to both climate change and health risks. These emissions further weaken community resilience by increasing the incidence of respiratory illness, particularly among children and the elderly.

7.4.6 Innovations and Community-Based Solutions

Despite these challenges, Southeast Asia has also been a site of grassroots innovation and community-driven responses. In Indonesia, the "Plastic Bank" initiative incentivizes community members to collect plastic waste in exchange for basic goods or digital tokens [63]. In the Philippines, barangays (villages) have piloted decentralized waste-segregation and composting schemes, often led by women's and youth organizations. These initiatives not only reduce plastic leakage but also foster local ownership and employment [64].

NGOs, such as GAIA Asia Pacific and Break Free From Plastic, have played an instrumental role in advancing zero-waste models and building local governments' capacity to implement circular economy principles. However, these programs are often under threat due to policy discontinuity and donor fatigue [65].

7.4.7 The Road Ahead: A Call for Systems Change

The challenges of waste infrastructure and microplastic management in Southeast Asia are deeply interlinked with structural inequalities, governance failures, and climate vulnerability. Tackling them requires a systems-thinking approach that includes:

- *Scaling up investment* in waste infrastructure through blended finance, climate funds, and green bonds.
- *Formalizing the informal sector* and ensuring labor rights, health access, and capacity building.
- *Strengthening policy coherence* at national and regional levels through ASEAN-led mechanisms.
- *Embedding environmental justice* in plastic policy to address the disproportionate burden borne by the poor.

Without decisive, inclusive, and well-funded interventions, Southeast Asia will continue to be a global hotspot of plastic pollution—threatening not only its rich biodiversity but also the well-being and dignity of its people.

7.5 Chapter Synthesis

This chapter demonstrates that microplastic pollution extends beyond environmental and health concerns into economic stability, social equity, and development pathways. The financial burden associated with plastic pollution continues to rise as governments allocate increasing resources for waste management, environmental monitoring, remediation programs, and public health protection. Coastal communities, tourism-dependent economies, and fisheries sectors face measurable losses linked to environmental degradation and declining ecosystem quality. These pressures highlight the economic urgency of preventive strategies and improved material management systems.

The analysis also shows that microplastic pollution intersects with climate processes. Plastic production relies heavily on fossil fuel feedstocks, while degradation processes contribute to greenhouse gas emissions throughout the plastic life cycle. Climate-related disasters further intensify plastic leakage through flood events, infrastructure damage, and disrupted waste collection systems. This interaction reveals a feedback relationship in which plastic pollution and climate change reinforce environmental vulnerability, particularly in coastal and climate-exposed regions.

Social dimensions further illustrate how exposure and risk remain unevenly distributed. Communities with limited waste infrastructure, informal recycling sectors, and high population density often experience greater environmental exposure. Small island states, coastal settlements, and rapidly urbanizing regions face structural constraints that limit effective waste recovery and pollution prevention. These conditions position microplastic pollution within the broader framework of environmental justice, where environmental risks often align with socioeconomic disadvantage.

Regional analysis highlights the distinct challenges faced by Southeast Asian countries. Rapid economic growth, increased plastic consumption, and gaps in waste management infrastructure contribute to high leakage rates into rivers and coastal waters. Informal waste sectors play an important role in material recovery, yet often operate without adequate occupational protection or environmental safeguards. Strengthening institutional coordination, investing in circular economy transitions, and supporting community-based waste solutions remain central to addressing these structural pressures.

The discussion, therefore, positions microplastic pollution as a development issue that requires integration across environmental governance, economic planning, climate adaptation, and social protection systems. Progress will depend on aligning scientific evidence with financial investments, governance reforms, and inclusive environmental management strategies that recognize regional realities and development priorities.

Economic pressures and social disparities highlight the limits of fragmented responses to plastic pollution. Effective solutions require coordinated governance frameworks, enforceable regulations, and international cooperation capable of addressing pollution across production, consumption, and waste management stages.

The next chapter, therefore, examines how policy instruments, regulatory frameworks, and global agreements shape responses to microplastic pollution. Attention turns toward international conventions, regional cooperation efforts, national regulatory strategies, and the institutional barriers that influence enforcement and effectiveness. This governance perspective connects scientific understanding with the policy mechanisms required to translate knowledge into measurable environmental progress.

References

1. Bank, M. S., & Hansson, S. V. (2019). The plastic cycle: A novel and holistic paradigm for the Anthropocene. *Environmental Science & Technology, 53*(13), 7177–7179. https://doi.org/10.1021/acs.est.9b02942
2. Lau, W. W. Y., Shiran, Y., Bailey, R. M., Cook, E., Stuchtey, M. R., Koskella, J., Velis, C. A., Godfrey, L., Boucher, J., Murphy, M. B., Thompson, R. C., Jankowska, E., Castillo Castillo, A., Pilditch, T. D., Dixon, B., Koerselman, L., Kosior, E., Favoino, E., Gutberlet, J., …, Palardy, J. E. (2020). Evaluating scenarios toward zero plastic pollution. *Science, 369*(6510), 1455–1461. https://doi.org/10.1126/science.aba9475
3. Ferronato, N., & Torretta, V. (2019). Waste mismanagement in developing countries: A review of global issues. *International Journal of Environmental Research and Public Health, 16*(6), 1060. https://doi.org/10.3390/ijerph16061060
4. Dauvergne, P. (2023). The necessity of justice for a fair, legitimate, and effective treaty on plastic pollution. *Marine Policy, 155*, 105785. https://doi.org/10.1016/j.marpol.2023.105785
5. Gallo, F., Fossi, M. C., Weber, R., Santillo, D., Sousa, J., Ingram, I., Nadal, Ã., & Romano, D. (2018). Marine litter plastics and microplastics and their toxic chemicals components: The need for urgent preventive measures. *Environmental Sciences, 30*(1). https://doi.org/10.1186/s12302-018-0139-z
6. OECD. (2023). Financing a future free from plastic leakage. In *OECD environment policy papers*. https://doi.org/10.1787/32424f03-en
7. Kibria, M. G., Masuk, N. I., Safayet, R., Nguyen, H. Q., & Mourshed, M. (2023). Plastic waste: Challenges and opportunities to mitigate pollution and effective management. *International Journal of Environmental Research, 17*(1), article 20. https://doi.org/10.1007/s41742-023-00507-z
8. Lavers, J. L., & Bond, A. L. (2017). Exceptional and rapid accumulation of anthropogenic debris on one of the world's most remote and pristine islands. *Proceedings of the National Academy of Sciences, 114*(23), 6052–6055. https://doi.org/10.1073/pnas.1619818114
9. Barboza, L. G. A., Dick Vethaak, A., Lavorante, B. R. B. O., Lundebye, A.-K., & Guilhermino, L. (2018). Marine microplastic debris: An emerging issue for food security, food safety and human health. *Marine Pollution Bulletin, 133*, 336–348. https://doi.org/10.1016/j.marpolbul.2018.05.047
10. Karasik, R., Lauer, N. E., Baker, A.-E., Lisi, N. E., Somarelli, J. A., Eward, W. C., Fürst, K., & Dunphy-Daly, M. M. (2023). Inequitable distribution of plastic benefits and burdens on economies and public health. *Frontiers in Marine Science, 9*. https://doi.org/10.3389/fmars.2022.1017247
11. Lebreton, L. C. M., van der Zwet, J., Damsteeg, J.-W., Slat, B., Andrady, A., & Reisser, J. (2017). River plastic emissions to the world's oceans. *Nature Communications, 8*, 15611. https://doi.org/10.1038/ncomms15611

12. Cordova, M. R., & Nurhati, I. S. (2019). Major sources and monthly variations in the release of land-derived marine debris from the greater Jakarta area, Indonesia. *Scientific Reports, 9*, 18730. https://doi.org/10.1038/s41598-019-55065-2
13. Keep America Beautiful. (2020). *2020 National Litter Study*.
14. Hanke, G., Ruiz-Orejon, L., Walvoort, D., & van Loon, W. (2025). Large scale trends of coastline litter including plastic as a proxy for impact mitigation in EU. In *One Ocean Science Congress 2025*. https://doi.org/10.5194/oos2025-1377
15. Mitrano, D. M., & Wohlleben, W. (2020). Microplastic regulation should be more precise to incentivize both innovation and environmental safety. *Nature Communications, 11*, 5324. https://doi.org/10.1038/s41467-020-19069-1
16. Abdul Aziz, M. F., & Raja Ariffin, R. N. (2024). The implementation of solid waste management policies in Southeast Asia: The present situations and future recommendations. *The Malaysian Journal of Social Administration, 18*, 110–132. https://doi.org/10.22452/18sp/113.159
17. Ratnawati, R. V., Tahar, N., Sidik, U. S., Onogawa, K., Gamaralalage, P. J. D., & Tsukiji, M. (2020). *National Plastic Waste Reduction Strategic Actions for Indonesia*. Ministry of Environment and Forestry. Retrieved from https://www.iges.or.jp/jp/pub/national-plastic-waste-reduction-strategic-actions-indonesia/en
18. Jang, Y. C., Hong, S., Lee, J., Lee, M. J., & Shim, W. J. (2014). Estimation of lost tourism revenue in Geoje Island from the 2011 marine debris pollution event in South Korea. *Marine Pollution Bulletin, 81*(1), 49–54. https://doi.org/10.1016/j.marpolbul.2014.02.021
19. UNEP. (2021). *From pollution to solution: A global assessment of marine litter and plastic pollution (synthesis report)*. United Nations Environment Programme. Retrieved from https://www.unep.org/resources/pollution-solution-global-assessment-marine-litter-and-plastic--pollution
20. Mejjad, N., El Mahdi Safni, A., & Laissaoui, A. (2025). Insightful analytical review of potential impacts of microplastic pollution on coastal and marine ecosystem services. *Journal of Hazardous Materials Advances, 17*, 100578. https://doi.org/10.1016/j.hazadv.2025.100578
21. Nyadjro, E. S., Cebrian, J., Cox, T. E., Wang, Z., Lau, Y., Konefal, A., Turnage, G., Offner, T., Gilpin, R., Boyer, T., Larsen, K., Mickle, P. F., Sparks, E., & Webster, J. A. B. (2026). Gaps and pathways towards standardized, FAIR microplastics data harmonization: A systematic review. *Microplastics, 5*(1), 11. https://doi.org/10.3390/microplastics5010011
22. Ruiz-Orejon, L., & Hanke, G. (2025). Marine microlitter and microplastic quality assurance and quality control—Towards a large scale approach. In *One Ocean Science Congress 2025*. https://doi.org/10.5194/oos2025-469
23. Wilson, D. C., Velis, C., & Cheeseman, C. (2006). Role of informal sector recycling in waste management in developing countries. *Habitat International, 30*(4), 797–808. https://doi.org/10.1016/j.habitatint.2005.09.005
24. Vince, J., & Hardesty, B. D. (2017). Plastic pollution challenges in marine and coastal environments: From local to global governance. *Restoration Ecology, 25*(1), 123–128. https://doi.org/10.1111/rec.12388
25. Lau, W. W. Y., Shiran, Y., Bailey, R. M., Cook, E., Stuchtey, M. R., Koskella, J., Velis, C. A., Godfrey, L., Boucher, J., Murphy, M. B., Thompson, R. C., Jankowska, E., Castillo Castillo, A., Pilditch, C. A., Dixon, B., Koerselman, L., Kosior, E., Favoino, E., Gutberlet, J., …, Palardy, J. E. (2020). *Breaking the plastic wave: A comprehensive assessment of pathways towards stopping ocean plastic pollution*. The Pew Charitable Trusts & SYSTEMIQ. Retrieved from https://www.pewtrusts.org/en/research-and-analysis/reports/2020/07/breaking-the-plastic-wave
26. Zheng, J., & Suh, S. (2019). Strategies to reduce the global carbon footprint of plastics. *Nature Climate Change, 9*, 374–378. https://doi.org/10.1038/s41558-019-0459-z
27. Geyer, R., Jambeck, J. R., & Law, K. L. (2017). Production, use, and fate of all plastics ever made. *Science Advances, 3*(7), e1700782. https://doi.org/10.1126/sciadv.1700782

28. Royer, S.-J., Ferrón, S., Wilson, S. T., & Karl, D. M. (2018). Production of methane and ethylene from plastic in the environment. *PLoS One, 13*(8), e0200574. https://doi.org/10.1371/journal.pone.0200574
29. Hamilton, L. A., & Feit, S. (2019). *Plastic & climate: The hidden costs of a plastic planet*. Center for International Environmental Law. Retrieved from https://www.ciel.org/plasticandclimate
30. Kvale, K., Hunt, C., James, A., & Koeve, W. (2023). Regionally disparate ecological responses to microplastic slowing of faecal pellets yields coherent carbon cycle response. *Frontiers in Marine Science, 10*, 1111838. https://doi.org/10.3389/fmars.2023.111
31. Cole, M., Lindeque, P., Fileman, E., Clark, J., Lewis, C., Halsband, C., & Galloway, T. S. (2015). The impact of polystyrene microplastics on feeding, function and fecundity in the marine copepod Calanus helgolandicus. *Environmental Science & Technology, 49*(2), 1130–1137. https://doi.org/10.1021/es504525u
32. Qayoom, I., Jan, F. Z., Rashid, I., Bhat, G. A., Ambili, A., & Sarangi, C. (2025). A review of microplastic contamination in the cryosphere. *iScience, 29*(1), 114414. https://doi.org/10.1016/j.isci.2025.114414
33. Zhu, X., Konik, J., & Kaufman, H. (2025). The knowns and unknowns in our understanding of how plastics impact climate change: A systematic review. *Frontiers in Environmental Science, 13*. https://doi.org/10.3389/fenvs.2025.1563488
34. Hopewell, J., Dvorak, R., & Kosior, E. (2009). Plastics recycling: Challenges and opportunities. *Philosophical Transactions of the Royal Society, B: Biological Sciences, 364*(1526), 2115–2126. https://doi.org/10.1098/rstb.2008.0311
35. de Souza Machado, A. A., Lau, C. W., Till, J., Kloas, W., Lehmann, A., Becker, R., & Rillig, M. C. (2018). Impacts of microplastics on the soil biophysical environment. *Environmental Science & Technology, 52*(17), 9656–9665. https://doi.org/10.1021/acs.est.8b02212
36. Corradini, F., Meza, P., Eguiluz, R., Casado, F., Huerta-Lwanga, E., & Geissen, V. (2019). Evidence of microplastic accumulation in agricultural soils from sewage sludge disposal. *Science of the Total Environment, 671*, 411–420. https://doi.org/10.1016/j.scitotenv.2019.03.368
37. Allen, S., Allen, D., Phoenix, V. R., Le Roux, G., Durántez Jiménez, P., Simonneau, A., Binet, S., & Galop, D. (2019). Atmospheric transport and deposition of microplastics in a remote mountain catchment. *Nature Geoscience, 12*, 339–344. https://doi.org/10.1038/s41561-019-0335-5
38. Bergmann, M., Mützel, S., Primpke, S., Tekman, M. B., Trachsel, J., & Gerdts, G. (2019). White and wonderful? Microplastics prevail in snow from the Alps to the Arctic. *Science Advances, 5*(8), eaax1157. https://doi.org/10.1126/sciadv.aax1157
39. UNEP. (2023). *Turning off the tap: How the world can end plastic pollution and create a circular economy*. United Nations Environment Programme. Retrieved from https://www.unep.org/resources/turning-off-tap
40. Hassen, M. B., Zouari, A. B., Abdennadher, M., Assaf, J. C., Nakad, M., Abboud, R., Khammeri, Y., Bannı, M., Panzeri, A., Gomes, L. A. d. V., & Hamd, W. (2025). Plastics pollution: Pathways, impacts, and regulatory challenges in marine environments. *Frontiers in Environmental Science, 13*. https://doi.org/10.3389/fenvs.2025.1635230
41. New, W. X., Kristanti, R. A., Manik, H., Wijayanti, Y., & Adeyemi, D. A. (2023). Occurrence of microplastics in drinking water in South East Asia: A short review. *Tropical Environment, Biology, and Technology, 1*(1), 14–24. https://doi.org/10.53623/tebt.v1i1.221
42. Chen, H. L., Selvam, S. B., Ting, K. N., & Gibbins, C. N. (2021). Microplastic pollution in freshwater systems in Southeast Asia: Contamination levels, sources, and ecological impacts. *Environmental Science and Pollution Research, 28*, 54222–54237. https://doi.org/10.1007/s11356-021-15746-8
43. Abahussain, A. A. M., Nasr, F. A., Jumah, A. B., Saravanan, P., Kumar, N. S., Al-Zharani, M., Guganathan, L., Sasikumar, G., Alsalamah, S. A., Qurtam, A. A., Kumar, J. S., & Tamizhdurai, P. (2025). Toxic threats from plastic waste: Human health impacts, challenges, and policy solutions. *RSC Advances, 15*(48), 40761. https://doi.org/10.1039/d5ra05845g
44. Sophie, R. W., Pacome, G. K. K., Aminata, G., Kelety, C., Barthélemy, T., Vinciale, A. A. W., Boubacar, N., & Ngangue, P. (2024). Impact of plastic waste on the human health in low-

income countries: A systematic review. *Journal of Environmental Protection, 15*(5), 572. https://doi.org/10.4236/jep.2024.155033

45. Ragusa, A., Svelato, A., Santacroce, C., Catalano, P., Notarstefano, V., Carnevali, O., Papa, F., Rongioletti, M. C. A., Baiocco, F., Draghi, S., D'Amore, E., Rinaldo, D., Matta, M., & Giorgini, E. (2021). Plasticenta: First evidence of microplastics in human placenta. *Environment International, 146*, 106274. https://doi.org/10.1016/j.envint.2020.106274
46. Wang, Y.-L., Lee, Y.-H., Chiu, I.-J., Lin, Y.-F., & Chiu, H.-W. (2020). Potent impact of plastic nanomaterials and micromaterials on the food chain and human health. *International Journal of Molecular Sciences, 21*(5), 1727. https://doi.org/10.3390/ijms21051727
47. Kadac-Czapska, K., Bukowska, B., Sicińska, P., & Grembecka, M. (2025). Hidden threats in infant diets and environment—Risks of microplastics and nanoplastics in food. *Reviews of Environmental Contamination and Toxicology, 263*(1). https://doi.org/10.1007/s44169-025-00099-z
48. Ubomba-Jaswa, E., & Kalebaila, N. (2020). Framing the plastic pollution problem within the water quality-health nexus: Current understandings and policy recommendations. *South African Journal of Science, 116*. https://doi.org/10.17159/sajs.2020/8115
49. Velis, C. A., & Cook, E. (2021). Mismanagement of plastic waste through open burning with emphasis on the global south: A systematic review of risks to occupational and public health. *Environmental Science & Technology, 55*(11), 7186. https://doi.org/10.1021/acs.est.0c08536
50. Cook, E., Derks, M., & Velis, C. A. (2023). Plastic waste reprocessing for circular economy: A systematic scoping review of risks to occupational and public health from legacy substances and extrusion. *Science of the Total Environment, 859*(Part 2), 160385. https://doi.org/10.1016/j.scitotenv.2022.160385
51. Zhang, Y., Kang, S., Allen, S., Allen, D., Gao, T., & Sillanpää, M. (2020). Atmospheric microplastics: A review on current status and perspectives. *Earth-Science Reviews, 203*, 103118. https://doi.org/10.1016/j.earscirev.2020.103118
52. Rochman, C. M., Tahir, A., Williams, S. L., Baxa, D. V., Lam, R., Miller, J. T., Teh, F.-C., Werorilangi, S., & Teh, S. J. (2015). Anthropogenic debris in seafood: Plastic debris and fibers from textiles in fish and bivalves sold for human consumption. *Scientific Reports, 5*, 14340. https://doi.org/10.1038/srep14340
53. Ng, C. H., Mistoh, M. A., Teo, S. H., Galassi, A., Ibrahim, A., Sipaut, C. S., Foo, J., Seay, J., Taufiq-Yap, Y. H., & Janaun, J. (2023). Plastic waste and microplastic issues in Southeast Asia. *Frontiers in Environmental Science, 11*, 1142071. https://doi.org/10.3389/fenvs.2023.1142071
54. Hardon, A., Dey, T., Pakasi, D., Taqueban, E. M., Hidayana, I., & Nichter, M. (2025). Confronting the material and structural leakiness of plastics: Insights from multi-sited ethnography in India, Indonesia and The Philippines. *Cambridge Prisms: Plastics, 3*, e6. https://doi.org/10.1017/plc.2025.10022
55. O'Hare, P., & Nøklebye, E. (2024). "The human face of the UN plastics treaty"? The role of waste pickers in intergovernmental negotiations to end plastic pollution and ensure a just transition. *Cambridge Prisms: Plastics, 2*. https://doi.org/10.1017/plc.2024.12
56. Marks, D. (2022). Transboundary governance failures and Southeast Asia's plastic pollution. In G. McCann & P. McCloskey (Eds.), *The Routledge handbook of global development* (1st ed.). Routledge. https://doi.org/10.4324/9781003009387
57. Telesetsky, A. (2023). The 'thin law' of plastic regulation and a proposal for a regional or global waste tariff. In F. M. Platjouw & A. Pozdnakova (Eds.), *The environmental rule of law for oceans: Designing legal solutions* (pp. 58–75). Cambridge University Press. https://doi.org/10.1017/9781009253746.007
58. Jesus, E. L. (2025). *Trends in addressing plastic pollution in The Philippines*. Asian Productivity Organization.
59. Sabatira, F. (2020). Southeast Asia regional cooperation on TACKLING marine plastic litter. *Lampung Journal of International Law, 2*(2), 69. https://doi.org/10.25041/lajil.v2i2.2033
60. Sumalde, Z. M. (2005). *Searching for a solution to solid waste in The Philippines (EEPSEA policy brief no. pb2005061)*. Economy and Environment Program for Southeast Asia (EEPSEA).

61. Nakajima, R., Miyama, T., Kitahashi, T., Isobe, N., Nagano, Y., Ikuta, T., Oguri, K., Tsuchiya, M., Yoshida, T., Aoki, K., Maeda, Y., Kawamura, K., Suzukawa, M., Yamauchi, T., Ritchie, H., Fujikura, K., & Yabuki, A. (2022). Plastic after an extreme storm: The typhoon-induced response of micro- and mesoplastics in coastal waters. *Frontiers in Marine Science, 8*. https://doi.org/10.3389/fmars.2021.806952
62. Kelly, F. J., Wright, S., Woodward, G., & Fussell, J. C. (2025). Plastic pollution under the influence of climate change: Implications for the abundance, distribution, and hazards in terrestrial and aquatic ecosystems. *Frontiers in Science, 3*. https://doi.org/10.3389/fsci.2025.1636665
63. Katz, D. (2019). Plastic Bank: Launching social plastic® revolution. *Field Actions Science Reports, Special Issue 19*, 96–99. Retrieved from https://journals.openedition.org/factsreports/5254
64. Alazada, A. M., & Magnaye, D. C. (2024). Greening the local solid waste management through community participation: Unfolding the challenges and creating opportunities for development planning. In *Building resilient and healthy cities: A guide to environmental sustainability and well-being* (pp. 7–24). Springer. https://doi.org/10.1007/978-981-99-7807-5_2
65. Makan, A., Salama, Y., Mamouni, F. Z., & Makan, M. (2025). Towards zero-waste cities: An integrated and circular approach to sustainable solid waste management. *Sustainability, 17*(17), 7884. https://doi.org/10.3390/su17177884

Part III
Taking Action

Chapter 8
Policies and Regulations

Introduction
Scientific evidence has clarified the scale, persistence, and transboundary movement of microplastics. Policy now determines whether knowledge translates into systemic change. The evolution of Extended Producer Responsibility, marine litter action plans, and regional cooperation mechanisms signals a shift from cleanup toward structural prevention. According to the Organisation for Economic Co-operation and Development, Extended Producer Responsibility extends a producer's accountability to the post-consumer stage of a product life cycle [1]. This principle underpins current legislative proposals in the Philippines and across Asia.

Asian developing countries have applied EPR to reduce marine plastic debris, though implementation varies by legal design and enforcement capacity [2]. Experience from India, Indonesia, and Japan demonstrates that regulatory clarity, accreditation of recyclers, and mandatory registration systems improve compliance and material recovery. In the Philippines, Senate Bill 2425 and House Bill 10696 institutionalize EPR for plastic packaging, amending Republic Act 9003. These measures align with the National Plan of Action for the Prevention, Reduction, and Management of Marine Litter [3].

Policy determines responsibility, financing, and accountability. Without enforceable frameworks, scientific findings remain advisory. With them, they reshape production, consumption, and waste governance.

8.1 International and Regional Frameworks: UN Treaties, ASEAN, and National Initiatives

International governance of plastic pollution now operates as a connected policy system rather than a set of isolated pledges. States first used existing agreements to manage downstream risks, especially waste trade and marine disposal. Negotiators

R. Y. Capangpangan, A. C. Alguno, *Microplastics in Focus*,
https://doi.org/10.1007/978-3-032-28978-0_8

now push upstream toward product design, chemicals of concern, production controls, and full lifecycle accountability. This shift reflects a practical lesson. Microplastic risk does not stay within borders, and end-of-pipe measures alone cannot keep plastics out of waterways, coasts, and food systems [4].

8.1.1 UN-Anchored Pathways: From Regime Complexity to a Life Cycle Instrument

The current policy landscape resembles a regime complex, with multiple treaties and institutions shaping different parts of the plastics problem. This structure offers immediate tools but also produces overlaps and gaps. For example, shipping and dumping rules limit direct inputs to the ocean, while chemicals and waste agreements regulate selected hazardous substances and transboundary movement. Yet microplastic leakage largely originates from land-based systems and product life cycles that these instruments only partly address [5].

Within this context, the UN process toward a global plastics instrument matters because it can consolidate fragmented obligations into a coherent framework. Legal analysis of the UNEA mandate highlights central debates that shape the emerging treaty architecture. Countries contest the scope, differentiation, financing, and the balance between harmonized global measures and nationally determined action. Crucially, they also debate whether the treaty should govern the full life cycle, including production and product design, or focus mainly on waste management and marine litter [4]. These disagreements determine whether the instrument can address microplastic drivers, such as additives, polymer choices, packaging intensity, and poor recyclability.

Treaty design also depends on participation and implementation. The negotiations increasingly involve non-state stakeholders, including industry and civil society, because they control key levers of change, such as material selection, packaging formats, and recovery investments. At the same time, governance scholars warn that stakeholder engagement must protect public interest and avoid capture, especially when negotiations seek controls on production and chemical composition [5].

8.1.2 Basel Convention Plastic Waste Controls: Tightening Trade as a Leakage Lever

Waste trade controls are among the strongest legally binding tools already in place for plastics governance. The Basel plastic waste amendments changed how countries classify and control shipments of plastic waste, expanding prior informed consent and raising scrutiny over mixed and contaminated streams. Evidence from trade and life-cycle assessment indicates that these amendments can reshape trade flows and environmental outcomes, especially by reducing shipments to destinations with a higher risk of mismanagement and open burning [6]. This matters for microplastics because mismanaged plastic waste fragments rapidly in the environment, then cycles into sediments, coasts, and biota.

However, tighter trade controls do not solve the upstream problem. They can shift burdens back to exporting states, which then need domestic collection, sorting, and end markets. In practice, Basel controls strengthen the case for circular economy policies at home, including design for recyclability and stable financing for recovery systems [6].

8.1.3 Extended Producer Responsibility as a Bridge Between Global Ambition and National Delivery

Extended Producer Responsibility, or EPR, has become a central implementation concept because it links upstream producers with downstream recovery outcomes. EPR also fits well within a life-cycle treaty logic because it assigns financial and organizational responsibilities across packaging value chains. Yet EPR performance depends on the quality of targets, monitoring, and enforcement. Empirical evaluation shows that EPR does not automatically reduce shoreline pollution unless policy design includes prevention-oriented metrics and monitoring systems aligned with the intervention [7]. This finding cautions against treating EPR as a label rather than a measurable accountability system.

Governance scholarship also positions EPR as a potential backbone for a future global agreement, especially if it drives design standards, transparent reporting, and predictable financing. At the same time, industry roles can either accelerate implementation through investment and compliance or weaken outcomes through minimal obligations and weak verification [8].

For developing contexts, design details matter even more. Evidence from Asian implementation indicates that EPR reduces marine plastic leakage when governments pair it with institutional capacity, clear reporting, and credible monitoring. Johannes et al. show that regulatory clarity and defined mechanisms improve transparency and recovery outcomes, while weak safeguards can shift costs to consumers and exclude vulnerable groups [2]. This insight aligns with policy design frameworks that recommend matching instruments to the impact pathway of plastics, from production and consumption to collection, treatment, and leakage points [9].

8.1.4 ASEAN: Soft Law Condition Under Real Structural Constraints

ASEAN's regional response relies heavily on coordination instruments rather than binding regional regulation. The region faces unique constraints that shape feasibility: archipelagic geographies, high transport costs, uneven infrastructure, and a large informal sector that already performs substantial recovery work. Comparative policy analysis across ASEAN countries highlights varied approaches, from bans and levies to voluntary programs, with uneven enforcement capacity and market readiness [10].

This ASEAN model still delivers value. It creates shared priorities, regional learning, and political signaling that can speed national reforms. It also supports donor and development programs that target municipal leakage points. A large multi-country program assessment in South and Southeast Asia shows that locally led initiatives can catalyze policy adoption, strengthen local government capacity, and support inclusion of informal collectors, but only when programs align with national plans and sustain financing beyond pilots [11]. For microplastics, this alignment matters because leakage often concentrates in rivers, drainage networks, and coastal hotspots rather than in national averages.

8.1.5 National Initiatives: The Philippine Case as an Implementation Laboratory

The Philippines illustrates why international and regional frameworks need strong domestic translation. Empirical assessments from the country show that food packaging and labels dominate litter in multiple environments, and riverbanks can act as high-density accumulation zones. These patterns reflect consumption systems based on small unit packaging and limited segregation capacity, which together increase fragmentation and microplastic formation [12].

In this context, EPR legislation and national marine litter planning serve as practical mechanisms for delivering on international commitments. The core design priorities remain consistent with global best practice. Set mandatory registration for obligated enterprises. Define recovery targets that rise over time. Require audited reporting. Accredit waste diverters. Build Producer Responsibility Organizations that can pool financing and standardize compliance. Most importantly, integrate informal sector safeguards so that reforms formalize protection and fair compensation rather than displacing livelihoods [2].

National success also depends on policy coherence. Basel-driven trade restrictions increase the need for domestic sorting and recycling markets. ASEAN coordination supports city-level pilots and harmonized messaging. A future UN treaty can supply global rules for product design, chemical management, and financing. Together, these levels can shift the system from cleanup to prevention by reducing unnecessary packaging, improving recyclability, and ensuring that producers pay for recovery rather than externalizing costs to municipalities and coastal communities [8, 9]. To better illustrate how global agreements, regional coordination platforms, and national regulatory systems interact to shape microplastic governance outcomes, Fig. 8.1 presents a conceptual multi-level governance framework. The figure highlights how international policy direction translates into regional coordination and national implementation, while cross-cutting stakeholders and enabling mechanisms support policy effectiveness across all governance levels.

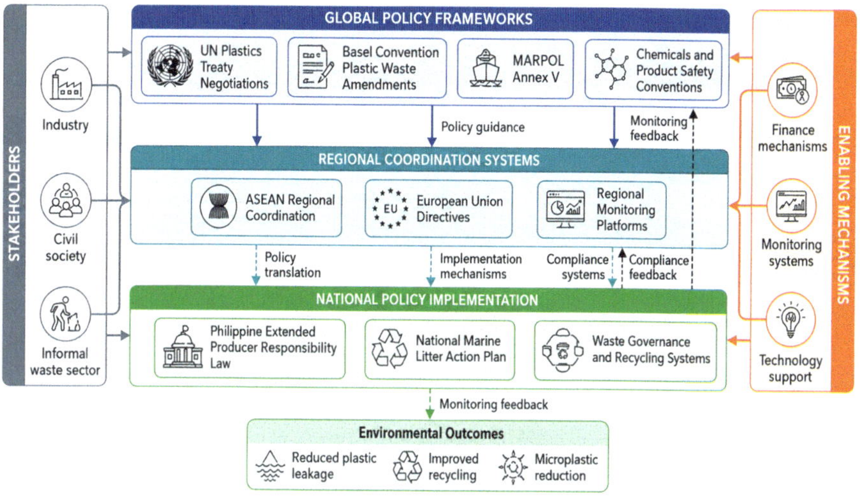

Fig. 8.1 Global microplastic policy governance framework. (Conceptual framework developed by the authors and illustrated using digital design tools)

8.2 Success Stories and Lessons Learned: Global South and Global North Comparisons

Policy succeeds when it changes everyday routines, not only statutes, and when scientific knowledge is effectively translated into regulatory action and measurable environmental outcomes (Fig. 8.2). Yet the route to those traits differs across the Global North and Global South because governance capacity, waste services, and market structure differ. A useful comparison does not rank countries. It explains why a policy tool works in one setting and how others can adapt it. The success of microplastic policies across different governance contexts can be understood through a common implementation pathway linking scientific evidence, policy design, operational mechanisms, and measurable environmental outcomes. Regardless of geographic context, effective policies typically follow a structured progression from evidence generation to regulatory instruments, implementation systems, behavioral change, and environmental improvement. Figure 8.2 illustrates the pathway through which scientific evidence is translated into policy instruments, implementation systems, behavioral change, and environmental outcomes, highlighting feedback mechanisms that support adaptive governance and policy effectiveness.

Importantly, this pathway also demonstrates that policy effectiveness depends on institutional coordination, sustained financing, and the capacity to monitor compliance across the value chain. Scientific data alone rarely produces change unless

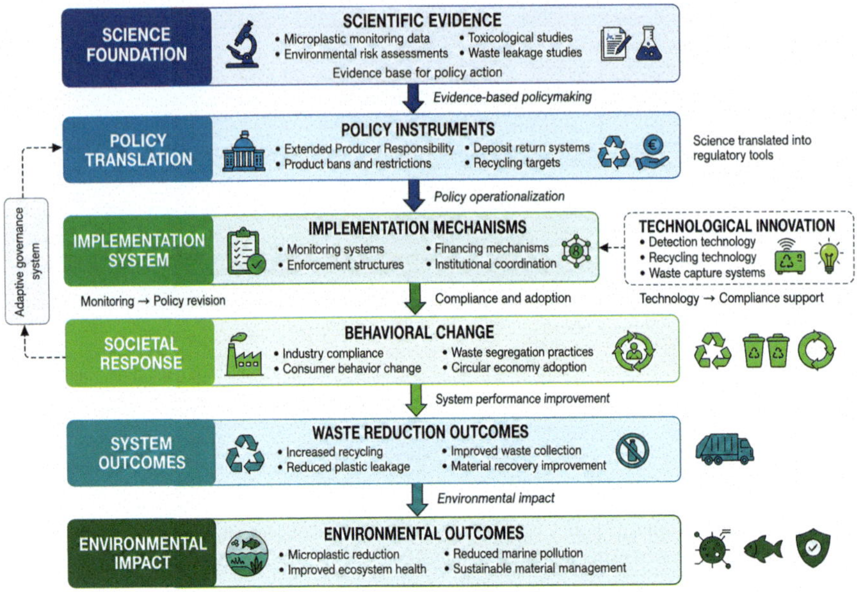

Fig. 8.2 Policy implementation pathway from science to environmental outcomes. (Conceptual framework developed by the authors and illustrated using digital design tools)

supported by enforcement mechanisms, stakeholder engagement, and incentives that encourage industry and consumer transitions toward sustainable practices. In many developing economies, community participation and informal sector integration are critical in bridging implementation gaps where formal waste systems remain limited. At the same time, continuous monitoring, data transparency, and impact evaluation enable policymakers to refine interventions and respond to emerging risks. These feedback loops strengthen regulatory learning and ensure that microplastic governance evolves alongside scientific advances and socioeconomic realities.

8.2.1 Global North Success Stories: High Compliance Through Clear Targets, Predictable Enforcement, and Monitoring

Microbead restrictions show what a narrow, science-backed ban can achieve. Regulators chose microbeads because they are intentionally added, easily defined, and replaceable. This design reduces industry resistance and improves compliance speed. Evidence from Toronto illustrates the payoff. After restrictions in Canada and the United States, researchers tracked microbeads in wastewater effluents and adjacent lake waters across multiple years and reported sharp declines, with reductions in effluents reaching up to 86%. This work matters because it links legislation to measurable environmental change, rather than assuming that a ban automatically reduces emissions [13].

Europe's microplastics story is more incremental. Several member states moved early through product standards and restrictions, while the EU advanced a broader restriction conversation on intentionally added microplastics in cosmetics. Kentin and Kaarto describe how the EU initiated a restriction procedure on microplastics in cosmetic products, showing how the EU often builds change through harmonization efforts that eventually standardize definitions and scope [14]. The lesson is practical. When regulators define the pollutant precisely and close loopholes, firms reformulate quickly because substitutes already exist. The microbead case also highlights the value of policy sequencing. Start with the simplest intentional source, then expand toward harder categories such as microfibers and tire wear.

Deposit return systems demonstrate how incentives and infrastructure combine. Germany's deposit return scheme illustrates how a deposit can normalize return behavior and raise collection rates for beverage containers. Rhein discusses lessons from the German experience and frames the scheme as a compliance system that delivers high return rates when retailers, reverse vending infrastructure, and clear obligations align [15]. The policy lesson extends beyond bottles. When a system makes the desired behavior convenient and rewarding, enforcement becomes less punitive because most people comply voluntarily.

The EU Single Use Plastics Directive shows that bans alone rarely deliver the full reduction. Kiessling and colleagues evaluated the directive's potential using large citizen science datasets and found that the largest reductions depend on implementation strength and complementary measures beyond bans, such as extended producer responsibility and product redesign requirements. In scenarios where only bans functioned, reductions remained modest. Where broader measures performed well, potential reductions rose substantially [16]. This is a key Global North lesson: advanced regulatory systems still need policy mixes. A ban addresses a visible item, but leakage persists unless redesign, collection, and financing close the loop.

Policy portfolio thinking explains why Global North frameworks often grow complex over time. A systematic analysis of plastics regulation across the European Union demonstrates that the number of policy instruments has increased substantially at both the EU and member state levels. However, the majority of measures concentrate on downstream stages of the plastics life cycle, particularly waste management and end-of-pipe controls. The study further shows that policymakers tend to rely more heavily on hierarchical instruments, such as standards, bans, and regulatory requirements, while market-based and informational instruments remain comparatively underutilized [17]. This pattern suggests that although Global North governance systems can sustain broader and more diversified policy portfolios, they may still overemphasize downstream control when upstream product design, material substitution, and consumption reduction strategies do not advance at the same pace.

8.2.2 *Global South Success Stories: Effectiveness Through State Credibility, Community Integration, and Adaptive Implementation*

Rwanda's plastic bag ban illustrates how consistent enforcement and norm shaping can drive rapid behavioral change. Rwanda implemented one of the strictest plastic bag bans globally, supported by border inspections, confiscation measures, and penalties. The effectiveness of the policy rests on a credible signal from the state that regulations will be applied uniformly, which accelerates social acceptance and compliance. Comparative political analysis shows that variation in plastic bag bans across countries often depends on enforcement credibility and political commitment. Rwanda represents a high-enforcement model in which sustained political will reduced resistance and strengthened compliance [18].

Although the ban directly targets macroplastics, its relevance to microplastics is clear. Plastic bags are a major source of secondary microplastics because they fragment under ultraviolet radiation, mechanical abrasion, and environmental weathering. As macroplastic inputs decline, downstream fragmentation into microplastics also decreases [19, 20]. Thus, Rwanda's macroplastic control contributes indirectly to microplastic mitigation by reducing environmental reservoirs that would otherwise degrade into smaller particles.

The Philippines and regional programs show why community systems matter when waste services remain uneven. Empirical assessments of plastic waste governance in South and Southeast Asia demonstrate that locally grounded initiatives achieve stronger outcomes when they combine public awareness, source segregation, and cross-sector partnerships. In a multi-country evaluation of plastic waste interventions in the Philippines, Vietnam, Sri Lanka, and Indonesia, locally led projects improved recovery performance when they worked through existing community networks and aligned local government, civil society organizations, and private recyclers [21]. These cases emphasized behavioral change campaigns, household segregation, and collaborative governance structures rather than relying solely on centralized enforcement.

A parallel body of research shows that integrating the informal waste sector is a decisive enabling condition in many Global South contexts. Studies on extended producer responsibility and recycling systems in Asian developing countries argue that informal waste collectors already perform a substantial share of collection and material recovery. When programs formalize, protect, and remunerate these actors, recovery rates increase and leakage declines [2]. Likewise, a global assessment of the informal recycling sector concludes that enabling informal workers through policy recognition, capacity building, and financial inclusion strengthens material circulation and reduces plastic leakage, particularly in cities where municipal collection coverage remains incomplete [22].

These lessons are especially visible in the Philippines and Vietnam, where projects supported independent waste picker organizations and strengthened downstream recycling value chains through partnerships and structured aggregation systems [21]. Such findings reflect structural realities across much of the Global South. Collection and sorting capacity often operates outside fully formalized municipal systems. Policy therefore performs better when it recognizes and invests in these decentralized actors rather than attempting to replace them abruptly. Strengthening community systems becomes a governance strategy grounded in institutional fit rather than institutional imitation.

Extended producer responsibility can work in the Global South when policy design fits market structure and directly supports leakage reduction. A peer-reviewed assessment of EPR implementation in Asian developing countries concludes that EPR contributes to reducing marine plastic debris when it aligns with existing waste management systems and strengthens collection and recycling channels. The authors caution against transplanting OECD models without adaptation to local institutional and economic realities, emphasizing the need to integrate informal recycling actors and tailor financing mechanisms to domestic conditions [2].

This insight is particularly relevant for the Philippines, where plastic consumption patterns include high volumes of low-value, multi-layer sachets, municipal waste management capacity varies widely across local government units, and informal waste collectors play a central role in material recovery [23]. Studies on plastic value chains in Southeast Asia show that effective plastic waste governance requires channeling financial resources toward collection and aggregation points and recognizing informal sector participation as part of the recovery system rather than treating it as an externality [22, 24]. When compliance mechanisms exclude micro enterprises and community recyclers, material leakage persists because collection networks remain underfunded or fragmented. Conversely, when EPR financing strengthens downstream sorting and recycling infrastructure while formalizing informal actors, recovery rates improve and environmental leakage declines [2, 22].

These findings support the broader argument that EPR effectiveness in the Global South depends less on formal policy adoption and more on institutional fit, inclusive financing, and recognition of decentralized recovery systems.

Informal sector inclusion is not a social add-on. It is a systems strategy grounded in empirical evidence. A global assessment of plastic pollution prevention argues that enabling the informal recycling sector offers a scalable pathway for low- and middle-income countries. Velis and colleagues show that reducing barriers to collection, improving remuneration and working conditions, and increasing material quality through source segregation and aggregation can significantly reduce plastic leakage. They further emphasize that interventions should prioritize plastic streams that generate the greatest environmental harm and leakage risk, thereby improving cost-effectiveness and environmental return per unit of investment [22].

These insights create a clear comparative lens showing that policy effectiveness depends not only on regulatory stringency or economic capacity, but on how well governance design aligns with institutional realities. In many Global North systems, waste governance relies heavily on formal compliance infrastructure, advanced sorting facilities, and producer responsibility organizations operating within regulated markets, often supported by strong monitoring capacity and regulatory depth. By contrast, in much of the Global South, collection and primary sorting already occur through informal networks, where policy success frequently emerges from institutional adaptation, community integration, and inclusive implementation strategies. Strengthening these existing actors through policy recognition, integration into extended producer responsibility (EPR) systems, and targeted financial support can deliver rapid leakage reduction while formal systems continue to mature. Evidence from integrated sustainable waste management research confirms that cities with strong informal sector integration often achieve higher recovery rates and lower leakage than those that attempt abrupt formalization without inclusion [25]. In this context, layering EPR financing and regulatory standards onto pre-existing collection systems can enhance traceability, improve material quality, and strengthen accountability without displacing livelihoods. The literature therefore supports a pragmatic sequencing approach: first stabilizing and professionalizing informal recovery systems, followed by aligning financial and regulatory instruments to reinforce rather than replace these networks [22]. Figure 8.3 synthesizes these contrasting governance pathways and highlights the shared lessons emerging across different development contexts.

The figure presents a comparative governance framework showing how Global North policies typically rely on strong regulatory systems and formal recycling infrastructure, while Global South approaches emphasize community integration, informal sector participation, and adaptive governance. It also highlights shared policy lessons demonstrating that effective microplastic governance depends on enforcement credibility, inclusive implementation, and sustainable financing, regardless of development context.

China's waste import ban reshaped global recycling incentives by abruptly restricting the flow of plastic waste into its domestic market. The National Sword policy, implemented in 2017–2018, set strict contamination thresholds and effectively halted imports of many categories of plastic waste. Empirical analyses demonstrate that this intervention disrupted long-established global recycling trade patterns, triggered price volatility, and redirected waste flows toward other Southeast Asian countries before many of them imposed similar restrictions [26, 27].

Xia and colleagues further show that the policy exposed structural weaknesses in exporting countries that had relied heavily on overseas processing capacity. The sudden contraction of Chinese import demand forced higher-income countries to confront deficiencies in domestic sorting, contamination control, and recycling infrastructure. In effect, the trade restriction functioned as an upstream intervention in the global plastics economy by shifting responsibility back to producers, consumers, and domestic waste management systems in exporting states [28].

Compared with China's National Sword policy, which illustrates how a single trade measure can influence material flows at scale. By altering market access rather than regulating products directly, the policy reconfigured incentives across supply chains. Export-dependent recycling systems were compelled to expand domestic processing capacity, reconsider product design for recyclability, and strengthen extended producer responsibility mechanisms to manage material within national borders [26–28].

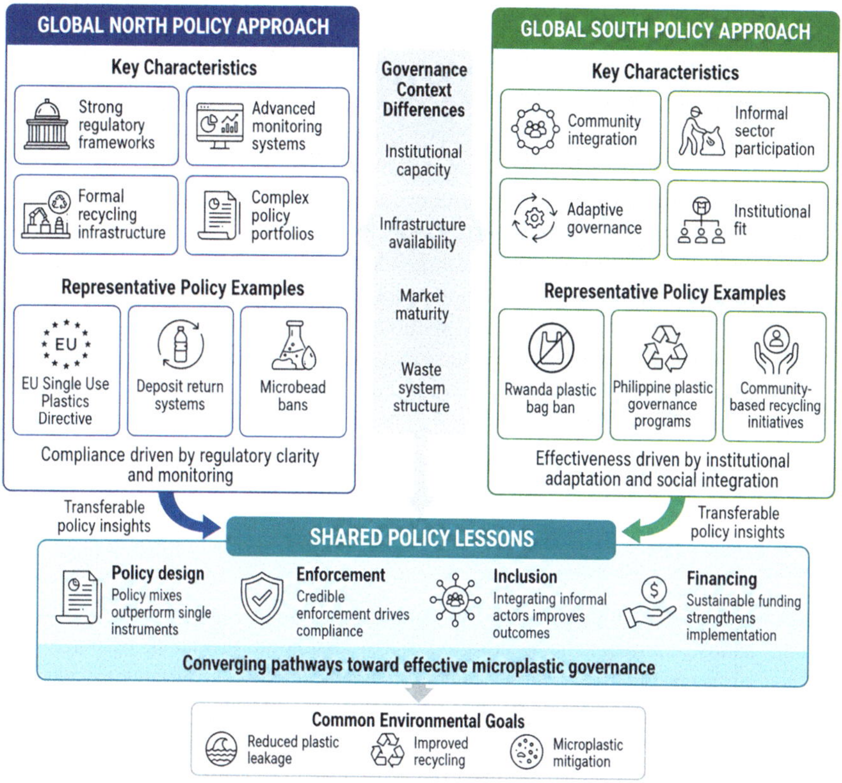

Fig. 8.3 Comparative policy effectiveness pathways in the Global North and Global South microplastic governance. (Conceptual framework developed by the authors and illustrated using digital design tools)

8.2.3 Cross-Cutting Lessons: What Travels Well Across North and South

1. *Start with a source that is easy to define and substitute, then scale outward.* Microbead bans succeeded because regulators targeted an intentional use that industry could replace quickly, and monitoring confirmed rapid declines [13].
2. *Do not treat enforcement as optional.* Rwanda shows that consistent enforcement builds credibility and fast norm change [18]. In the EU context, portfolio growth and directive design still depend on implementation quality at national and local levels.
3. *Pair rules with a practical compliance pathway.* Local initiatives in the Philippines and Vietnam show that awareness, segregation, and partnerships support compliance by making the desired behavior feasible.

4. *Treat informal recyclers as infrastructure*. Formalizing and supporting informal collectors mobilizes an existing workforce, thereby improving recovery and reducing leakage, especially in lower-income contexts [22].
5. *Use policy mixes, not single tools*. Evaluations of the EU Single Use Plastics Directive show that bans alone achieve limited reductions, whereas combining extended producer responsibility, product redesign, improved collection, and economic and informational instruments across the life cycle delivers substantially greater leakage reduction [16, 17].
6. *Build resilience into implementation*. Evidence from Southeast and South Asia shows that plastic waste governance must remain adaptive to shocks, such as pandemics and market disruptions, with resilient systems relying on decentralized leadership, community partnerships, diversified financing, and sustained engagement with informal actors to maintain continuity and long-term pollution control [21, 22].

8.3 Policy Gaps and Enforcement Barriers

Despite rapid growth in global plastics regulation, implementation gaps remain significant. Many countries have adopted bans, extended producer responsibility schemes, and marine litter action plans, yet plastic leakage continues to rise. The challenge no longer lies in policy absence alone. It lies in fragmented governance, weak enforcement infrastructure, limited financing, transboundary waste trade, and inadequate monitoring harmonization. These structural barriers slow the translation of legislative ambition into measurable environmental outcomes.

8.3.1 Fragmented Governance and Institutional Overlap

Plastic governance often spans multiple ministries, including environment, trade, industry, fisheries, customs, and local government. This dispersion creates coordination gaps and unclear accountability. Liu (2025) [29] identifies fragmented institutional mandates as a primary barrier to effective plastic policy implementation, noting that marine plastic governance frequently operates across disconnected regulatory silos. Their analysis demonstrates that when responsibilities overlap without integrated coordination mechanisms, enforcement weakens and compliance declines [29].

This fragmentation is especially visible in extended producer responsibility systems. While EPR requires collaboration between environmental agencies and industry regulators, misaligned mandates can delay the accreditation of producer responsibility organizations, weaken reporting systems, and reduce clarity in enforcement [2]. Without centralized oversight and cross-agency coordination platforms, regulatory expansion increases complexity without proportionate compliance gains.

8.3.2 *Insufficient Monitoring and Technical Infrastructure*

Effective enforcement depends on measurable indicators. Yet monitoring infrastructure across many regions remains inadequate. Hassan et al. (2024) [30] document substantial gaps in marine debris monitoring capacity in ASEAN countries, including inconsistent methodologies, limited laboratory capability for microplastic analysis, and fragmented reporting systems. The authors emphasize that without standardized monitoring frameworks, national reduction targets cannot be validated, and policy impact remains uncertain [30].

Globally, methodological inconsistencies in microplastic sampling, extraction, and identification further complicate data comparability. A comprehensive review by Cowger et al. (2021) highlights the absence of harmonized laboratory protocols as a major obstacle to reliable microplastic quantification. Without standardized approaches, reported reduction trends may reflect methodological variation rather than genuine environmental improvement [31].

The implication is direct. Enforcement agencies cannot sanction non-compliance or reward progress without credible data. Policy evaluation thus depends on investment in laboratory infrastructure, standardized reporting sheets, and digital traceability systems that connect collection, recycling, and disposal records.

8.3.3 *Funding Constraints and Enforcement Capacity*

Legislation without financial backing yields limited results. UNEP estimated that the costs of environmental damage associated with plastic pollution exceed USD 13 billion annually, reflecting impacts on marine ecosystems, fisheries, tourism, and coastal infrastructure [32]. Despite such losses, waste management budgets in many low- and middle-income countries remain constrained.

Extended producer responsibility offers one financing pathway, yet implementation requires administrative systems, auditing mechanisms, and enforcement personnel. Research on EPR in Asian developing countries shows that insufficient funding and weak cost recovery systems constrain compliance monitoring and reporting [2]. Where local governments lack technical and financial capacity, enforcement becomes reactive rather than preventive.

Even in higher-income countries, underfunded inspection regimes limit regulatory effectiveness. Without consistent auditing of producer declarations and recovery claims, leakage persists despite formal compliance.

8.3.4 Transboundary Plastic Waste Trade

International trade complicates domestic regulation. Lebreton and Andrady (2019) demonstrate that global plastic waste flows redistribute environmental burdens from exporting to importing countries, often concentrating processing in regions with weaker environmental safeguards. Such trade patterns obscure accountability because leakage may occur far from the point of production [33].

China's National Sword policy illustrated how trade restrictions can abruptly expose domestic waste management weaknesses in exporting countries [26]. When import pathways closed, several countries struggled to process accumulated waste domestically, revealing overreliance on external processing markets.

Transboundary waste flows therefore represent both a governance gap and an enforcement challenge. Customs inspections, contamination thresholds, and Basel Convention amendments attempt to regulate trade, yet illegal shipments and misclassification remain enforcement obstacles. Effective plastic governance requires traceability mechanisms that extend beyond national borders.

8.3.5 Informal Sector Exclusion

In many developing contexts, informal waste collectors recover a substantial share of recyclable material. Yet formal regulatory frameworks often overlook or marginalize these actors. Velis et al. (2022) [22] argue that excluding informal recyclers from policy design weakens recovery systems and increases leakage risk. When enforcement focuses solely on formal operators, material handled by informal actors goes undocumented, reducing data accuracy and oversight of compliance.

Inclusive governance, therefore, serves both social and enforcement objectives. Integrating informal networks into EPR reporting systems enhances traceability and improves transparency of material flows.

8.3.6 Data Harmonization and Performance Metrics

One of the most persistent policy gaps lies in data harmonization. Countries report plastic waste generation, recycling rates, and leakage using different definitions and calculation methods. This inconsistency complicates global benchmarking and obscures progress assessment.

Cowger et al. (2021) [31] emphasize that standardized protocols for measuring microplastics are essential for evaluating policy effectiveness. Without methodological alignment, trend analyses remain unreliable. Similarly, global assessments of marine plastic pathways stress the need for unified indicators across land-based and marine monitoring systems [34].

The absence of harmonized data frameworks undermines accountability mechanisms embedded in national action plans and international negotiations. A reduction target cannot be verified if baseline conditions vary across reporting systems.

8.3.7 *Political Economy and Compliance Incentives*

Beyond technical barriers, enforcement often encounters political economy constraints. The plastic production and packaging industries are significant economic sectors. Regulatory expansion may face resistance when compliance costs threaten competitiveness.

Eckert et al. (2024) [17] show that while European plastic policy portfolios have expanded, many instruments still emphasize downstream waste management rather than upstream production controls. This pattern reflects political feasibility dynamics, where bans and standards gain support more readily than systemic production limits. Such imbalances may reduce transformative potential.

Effective enforcement of plastic governance increasingly depends on aligning economic incentives with environmental objectives, since regulatory bans alone are often undermined by market conditions that favor cheap virgin plastics and weak compliance. As shown in Fig. 8.4, the core challenge is no longer the absence of policies but the difficulty of translating policy commitments into effective implementation. This implementation gap arises from fragmented institutional mandates, weak inter-agency coordination, limited monitoring capacity, insufficient financing, and inadequate technical infrastructure. Additional complications emerge from transboundary plastic waste trade, weak traceability systems, and the exclusion of informal waste actors, all

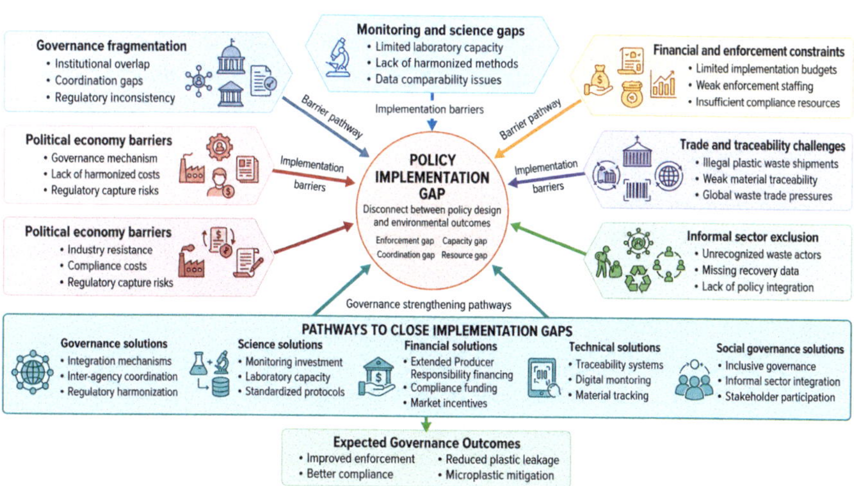

Fig. 8.4 Policy gaps and enforcement barriers affecting microplastic governance. (Conceptual framework developed by the authors and illustrated using digital design tools)

of which reduce enforcement effectiveness and obscure actual material flows. Political economy pressures, including industry resistance and compliance cost concerns, further demonstrate that implementation failures are shaped not only by technical limitations but also by governance and economic realities.

Addressing these persistent gaps requires integrated governance strategies that move beyond isolated regulatory approaches. As synthesized in the figure, governance strengthening pathways include regulatory harmonization, scientific capacity building, sustainable financing mechanisms such as extended producer responsibility, and the deployment of digital traceability and monitoring systems. Equally important are inclusive governance approaches that recognize the role of informal waste sectors and promote multi-stakeholder participation. Together, these coordinated interventions help close implementation gaps by improving enforcement capacity, strengthening compliance systems, reducing plastic leakage, and advancing microplastic mitigation. Figure 8.4, therefore, highlights an important shift in plastic governance: environmental success depends less on the number of policies adopted and more on the strength, coordination, and inclusiveness of the systems that enable their effective implementation.

8.4 Annex: Global and Regional Policy Tools

A. **Global Policy Tools**

1. Binding international instruments and rules
   - *Global plastics treaty instrument* (treaty architecture tool): harmonized definitions, phased controls, national action plans, reporting, compliance committee, and finance mechanism.
   - *Basel Convention plastic waste controls*: prior informed consent for transboundary shipments, tighter controls on mixed and contaminated plastics to reduce waste dumping risks.
   - *MARPOL Annex V* (shipping): strengthened port reception facilities, reporting of lost cargo, and enforcement of discharge prohibitions.
   - *Chemicals and product safety conventions* (where relevant to additives): manage hazardous additives that complicate recycling and create health risks.
2. Global standards, harmonization, and measurement
   - *Standardized monitoring protocols* for microplastics (sampling, QA/QC, polymer identification, and reporting templates) to enable comparability and verify reduction trends [31].
   - *Life cycle-based policy portfolio mapping* to identify gaps upstream vs. downstream and balance instrument mix [17].
3. Global finance and market-shaping tools

- *International finance windows* for collection, sorting, safe disposal, and wastewater upgrades (especially where leakage is highest).
- *Trade and supply chain tools*: transparency on plastic feedstock, resin and packaging flows; minimum recycled content benchmarks; restrictions on problematic products in international supply chains.

B. **Regional Policy Tools**

1. European Union and EEA-style tools

- *Directive-based controls* (bans, design requirements, EPR, and collection targets) with enforceable national transposition.
- *Policy mixes* rather than bans alone: evidence from EU assessments shows higher reduction potential when bans are paired with EPR, redesign obligations, and collection improvements [16, 17].

2. ASEAN and wider Asia Pacific tools

- *Regional monitoring and reporting platforms* to reduce method fragmentation and strengthen comparability across coastlines and river basins [35].
- *EPR "fit-for-context" frameworks* that integrate informal recovery systems and strengthen collection and recycling channels rather than copying OECD models without adaptation [2].

3. Africa and small island regions tools

- *High-enforcement product bans/controls* (bags, selected single-use items) plus border control and retailer compliance systems.
- *Regional procurement standards* to shift public purchasing away from high-leakage plastics.
- *Island-focused leakage controls*: port waste reception, fishing gear management, and stormwater litter interception.

C. **Cross-cutting categories you can apply anywhere**

1. Regulatory tools

- *Product bans and restrictions* (microbeads; selected single-use items).
- *Performance standards* (recycled content, recyclability, durability).
- *Permit and discharge controls* for industrial pellets, wastewater plants, and stormwater.

2. Economic tools

- *EPR fees modulated by design* (higher fees for multi-layer, non-recyclable formats).
- *Deposit return systems* for high-value containers.
- *Levies/taxes* on virgin resin or problematic packaging.

3. Infrastructure and enforcement tools

- *Inspection and audit systems* for EPR claims and recovery certificates.

- *Border controls and contamination thresholds* for imported waste and materials.
- *Wastewater and stormwater upgrades* (filters, screening, capture devices).

4. Information tools

- *Labeling and disclosure* (polymer type, recycled content, recyclability).
- *Digital product passports* and traceability for packaging and pellets.
- *Public reporting dashboards* tied to verified metrics.

5. Inclusion tools

- *Formal recognition and contracting* of informal waste collectors as part of recovery systems, improving capture and data quality [22].

8.5 Chapter Synthesis

This chapter establishes that effective responses to microplastic pollution depend on the strength of governance systems that translate scientific knowledge into enforceable action. International agreements, regional coordination platforms, and national regulatory frameworks increasingly recognize the need to address plastics across their full life cycle, from material design and production to waste recovery and environmental protection. These governance developments reflect a shift from reactive cleanup approaches toward preventive strategies that target the structural drivers of plastic leakage.

Comparative policy experience demonstrates that regulatory success depends on alignment between policy design, enforcement capacity, and local institutional realities. Experiences from the Global North highlight the importance of clear regulatory targets, monitoring systems, and policy portfolios that combine bans, extended producer responsibility, and product redesign requirements. In contrast, Global South experiences emphasize the value of adaptive governance, community integration, and institutional fit, particularly where informal recovery systems already perform critical waste management functions. These comparative insights show that effective policy does not rely on uniform models but on context-responsive implementation.

This chapter also highlights persistent governance gaps that limit the effectiveness of policies. Fragmented institutional mandates, insufficient monitoring infrastructure, funding constraints, and inconsistencies in data reporting continue to slow progress in many jurisdictions. Transboundary plastic waste trade further complicates national enforcement by redistributing environmental burdens across borders. Addressing these barriers requires coordinated regulatory approaches, investment in monitoring capacity, harmonized reporting frameworks, and financing systems that sustain long-term compliance.

Policy effectiveness also depends on inclusive governance. Evidence presented in this chapter shows that integrating informal waste collectors, strengthening producer accountability, and aligning economic incentives with environmental objectives can significantly improve recovery outcomes and reduce leakage. These approaches demonstrate that plastic governance functions most effectively when environmental protection aligns with economic opportunity and social inclusion.

The discussion, therefore, positions policy as a central mechanism for systemic change. Regulatory frameworks provide direction, but their effectiveness depends on the quality of implementation, scientific support, technological feasibility, and stakeholder engagement. Progress in addressing microplastic pollution will depend on governance systems that combine regulatory clarity, scientific evidence, financial mechanisms, and institutional coordination to sustain long-term environmental improvements.

Policy frameworks define direction and responsibility, yet their success depends on the availability of practical solutions that enable compliance and accelerate environmental improvement. Regulatory ambition requires technological pathways that support material substitution, improve recycling efficiency, strengthen waste capture systems, and reduce environmental leakage across production and consumption systems.

The next chapter, therefore, examines scientific and technological innovations that support the transition toward sustainable materials management. Attention turns to advances in biodegradable materials, circular-economy strategies, improved recycling technologies, and emerging cleanup solutions that translate policy commitments into operational progress. This perspective highlights the role of innovation as a critical bridge between regulatory intent and measurable environmental outcomes.

References

1. Organisation for Economic Co-Operation and Development (OECD). (2024). *Extended producer responsibility: Basic facts and key principles* (OECD Environment Policy Paper No. 41). OECD Publishing.
2. Johannes, H. P., Kojima, M., Iwasaki, F., & Edita, E. P. (2021). Applying the extended producer responsibility towards plastic waste in Asian developing countries for reducing marine plastic debris. *Waste Management & Research: The Journal for a Sustainable Circular Economy, 39*(5), 690–702. https://doi.org/10.1177/0734242X211013412
3. Rosenberg, M. (2022). Philippines. *Asia-Pacific Journal of Ocean Law and Policy, 7*(2), 300–306. https://doi.org/10.1163/24519391-07020008
4. Wang, S. (2023). International law-making process of combating plastic pollution: Status quo, debates and prospects. *Marine Policy, 147*, 105376. https://doi.org/10.1016/j.marpol.2022.105376
5. Stöfen-O'Brien, A. (2025). Negotiating plastics futures: The Law of the sea and the role of non-state stakeholders. *International and Comparative Law Quarterly, 74*(S1), 291–303. https://doi.org/10.1017/S002058932510119X
6. Li, K., Ward, H., Lin, H. X., & Tukker, A. (2025). Shift to intra-EU-OECD trade enhanced environmental benefits after Basel convention plastic waste amendments. *Resources, Conservation and Recycling, 223*, 108527. https://doi.org/10.1016/j.resconrec.2025.108527

7. Harris, L., et al. (2021). Using citizen science to evaluate extended producer responsibility policy to reduce marine plastic debris shows no reduction in pollution levels. *Marine Policy, 123*, 104319. https://doi.org/10.1016/j.marpol.2020.104319
8. Raubenheimer, K., & Urho, N. (2020). Rethinking global governance of plastics: The role of industry. *Marine Policy, 113*, 103802. https://doi.org/10.1016/j.marpol.2019.103802
9. Alpizar, F., Carlsson, F., Lanza, G., Carney, B., Daniels, R. C., Jaime, M., Ho, T., Nie, Z., Salazar, C., Tibesigwa, B., & Wahdera, S. (2020). A framework for selecting and designing policies to reduce marine plastic pollution in developing countries. *Environmental Science & Policy, 109*, 25–35. https://doi.org/10.1016/j.envsci.2020.04.007
10. Hermawan, S., & Astuti, W. (2021). Analysing several ASEAN countries' policy for combating marine plastic litter. *Environmental Law Review, 23*(1), 55–73. https://doi.org/10.1177/146145292
11. Mathis, J. E., Gillet, M. C., Disselkoen, H., & Jambeck, J. R. (2022). Reducing ocean plastic pollution: Locally led initiatives catalyzing change in south and Southeast Asia. *Marine Policy, 143*, 105127. https://doi.org/10.1016/j.marpol.2022.105127
12. Fuentes, H., Flores, J. P., Bacaltos, D. G., Fuentes, A., Roman, L., Schuyler, Q., & Hardesty, B. D. (2024). A comprehensive assessment of plastic pollution inland, on riverbanks, along the coastline and at the sea surface in southern Mindanao, Philippines. *Marine Pollution Bulletin, 209*(Part B), 117134. https://doi.org/10.1016/j.marpolbul
13. Akhbarizadeh, R., Yu, J. I., Ead, L., Nicholls, E., Thibeau, J., Lanisa, M., Wakai, M., Marquez, A., Miller, C., Sims, A., Diamond, M. L., & Helm, P. A. (2024). Reductions of plastic microbeads from personal care products in wastewater effluents and lake waters following regulatory actions. *ACS ES&T Water, 4*(2), 492–499. https://doi.org/10.1021/acsestwater
14. Kentin, E., & Kaarto, H. (2018). An EU ban on microplastics in cosmetic products and the right to regulate. *Review of European, Comparative & International Environmental Law, 27*(3), 254–266. https://doi.org/10.1111/reel.12269
15. Rhein, S., & Sträter, K. F. (2021). Intended and unintended effects of statutory deposit return schemes for single-use plastic bottles: Lessons learned from the German experience. *GAIA—Ecological perspectives for Science and Society, 30*(4), 250–256. https://doi.org/10.14512/gaia.30.4.8
16. Kiessling, T., Hinzmann, M., Medereke, L., Dittmann, S., Brennecke, D., Böhm-Beck, M., Nickmeier, K., & Thiel, M. (2023). What potential does the EU single-use plastics directive have for reducing plastic pollution at coastlines and riversides? An evaluation based on citizen science data. *Waste Management, 164*, 106–118. https://doi.org/10.1016/j.wasman.2023.03.023
17. Eckert, S., Karassin, O., & Steinbach, Y. (2024). A policy portfolio approach to plastics throughout their life cycle: Supranational and national regulation in the European Union. *Environmental Policy and Governance, 34*(4), 427–441. https://doi.org/10.1002/eet.2092
18. Behuria, P. (2021). Ban the (plastic) bag? Explaining variation in the implementation of plastic bag bans in Rwanda, Kenya and Uganda. *Environment and Planning C: Politics and Spaces, 39*(8), 1791–1808. https://doi.org/10.1177/2399654421994836
19. Andrady, A. L. (2011). Microplastics in the marine environment. *Marine Pollution Bulletin, 62*(8), 1596–1605. https://doi.org/10.1016/j.marpolbul.2011.05.030
20. Barnes, D. K. A., Galgani, F., Thompson, R. C., & Barlaz, M. (2009). Accumulation and fragmentation of plastic debris in global environments. *Philosophical Transactions of the Royal Society, B: Biological Sciences, 364*(1526), 1985–1998. https://doi.org/10.1098/rstb.2008.0205
21. Benson, N. U., Bassey, D. E., & Palanisami, T. (2021). COVID pollution: Impact of COVID-19 pandemic on global plastic waste footprint. *Heliyon, 7*(2), e06343. https://doi.org/10.1016/j.heliyon.2021.e06343
22. Velis, C. A., Hardesty, B. D., Cottom, J. W., & Wilcox, C. (2022). Enabling the informal recycling sector to prevent plastic pollution and deliver an inclusive circular economy. *Environmental Science & Policy, 138*, 20–25. https://doi.org/10.1016/j.envsci.2022.09.008
23. Mihai, F. C., Gündoğdu, S., Markley, L., Olivelli, A., Khan, F. R., Gwinnett, C., Gutberlet, J., Reyna-Bensusan, N., Llanquileo-Melgarejo, P., Meidiana, C., Elagroudy, S., Ishchenko, V.,

Penney, S., Lenkiewicz, Z., & Molinos-Senante, M. (2021). Plastic pollution, waste management issues, and circular economy opportunities in rural communities. *Sustainability, 14*(1), 20. https://doi.org/10.3390/su14010020
24. Gutberlet, J. (2023). Global plastic pollution and informal waste pickers. *Cambridge Prisms: Plastics, 1*. https://doi.org/10.1017/plc.2023.10
25. Wilson, D. C., Velis, C., & Cheeseman, C. (2006). Role of informal sector recycling in waste management in developing countries. *Habitat International, 30*(4), 797–808. https://doi.org/10.1016/j.habitatint.2005.09.005
26. Brooks, A. L., Wang, S., & Jambeck, J. R. (2018). The Chinese import ban and its impact on global plastic waste trade. *Science Advances, 4*(6), eaat0131. https://doi.org/10.1126/sciadv.aat0131
27. Wang, C., Zhao, L., Lim, M. K., Chen, W.-Q., & Sutherland, J. W. (2020). Structure of the global plastic waste trade network and the impact of China's import ban. *Resources, Conservation and Recycling, 153*, 104591. https://doi.org/10.1016/j.resconrec.2019.104591
28. Li, B., & Mu, Y. (2024). Impact of China's National Sword policy on waste import margins: A difference-in-differences approach. *Sustainability, 16*(2), 776. https://doi.org/10.3390/su16020776
29. Liu, Z. (2025). A maritime community with a shared future: Bridging fragmented legal regulations for marine plastic pollution. *Frontiers in Marine Science, 12*, 1680110. https://doi.org/10.3389/fmars.2025.1680110
30. Hassan, I., Sethupathi, S., Bashir, M. J. K., Munusamy, Y., & Chan, C. W. (2024). A systematic review of microplastics occurrence, characteristics, identification techniques and removal methods in ASEAN and its future prospects. *Journal of Environmental Chemical Engineering, 12*(2), 112305. https://doi.org/10.1016/j.jece.2024.112305
31. Cowger, W., Steinmetz, Z., Gray, A., Munno, K., Lynch, J., Hapich, H., Primpke, S., De Frond, H., Rochman, C., & Herodotou, O. (2021). Microplastic spectral classification needs an open source community: Open Specy to the rescue! *Analytical Chemistry, 93*(21), 7543–7548. https://doi.org/10.1021/acs.analchem.1c00123
32. United Nations Environment Programme. (2014, June 23). *Plastic waste causes financial damage of US$13 billion to marine ecosystems each year as concern grows over microplastics*. Retrieved from https://www.unep.org/news-and-stories/press-release/plastic-waste-causes-financial-damage-us13-billion-marine-ecosystems
33. Lebreton, L., & Andrady, A. (2019). Future scenarios of global plastic waste generation and disposal. *Palgrave Communications, 5*, 6. https://doi.org/10.1057/s41599-018-0212-7
34. Jambeck, J. R., Geyer, R., Wilcox, C., Siegler, T. R., Perryman, M., Andrady, A., Narayan, R., & Law, K. L. (2015). Marine pollution. Plastic waste inputs from land into the ocean. *Science, 347*(6223), 768–771. https://doi.org/10.1126/science.1260352
35. Bakir, A., McGoran, A. R., Silburn, B., Russell, J., Nel, H., Lusher, A. L., Amos, R., Shadrack, R. S., Arnold, S. J., Castillo, C., Urbina, J. E., Barrientos, E., Sanchez, H., Pillay, K., Human, L., Swartbooi, T., Cordova, M. R., Sani, S. Y., Wijesinghe, T. W. A. W., …, Mayes, A. G. (2024). Creation of an international laboratory network towards global microplastics monitoring harmonisation. *Scientific Reports, 14*(1), 12714. https://doi.org/10.1038/s41598-024-62176-y

Chapter 9
Technological and Scientific Innovations

Introduction

Science has clarified the scale of microplastic pollution. Action now depends on innovation. This chapter presents technological and scientific advances that reshape how societies design, use, recover, and replace plastics. It draws from recent evidence on biodegradable polymers in agriculture, additive risks, environmental degradation pathways, and soil ecosystem responses [1–3].

Technological innovation operates across the plastic life cycle. It begins with material redesign through bio-based and biodegradable polymers. It advances through circular economy strategies that reduce waste generation and extend the value of materials. It integrates improved waste management systems into emerging economies. It expands to environmental cleanup technologies and sector-specific mitigation in agriculture and aquaculture.

However, innovation must align with environmental reality. Biodegradability depends on system properties that combine material chemistry and environmental conditions [1, 4]. Laboratory certification does not guarantee field performance. Fragmentation into micro-bioplastics may still occur before mineralization [5, 6]. Technological progress, therefore, requires scientific rigor, transparent standards, and long-term field validation. Only then can innovation reduce microplastic burdens rather than shift them across ecosystems.

R. Y. Capangpangan, A. C. Alguno, *Microplastics in Focus*,
https://doi.org/10.1007/978-3-032-28978-0_9

9.1 Bioplastics, Circular Economy, and Sustainable Packaging

9.1.1 *Rethinking Plastic at the Molecular Level*

Bioplastics have emerged as a strategic response to persistent plastic pollution. Yet, terminology often confuses policymakers, producers, and consumers. Bio-based materials derive from renewable feedstocks, such as corn, sugarcane, or cellulose. Biodegradable materials decompose through microbial processes into water, carbon dioxide, methane, and biomass under specific environmental conditions [1]. These categories do not always overlap. Biopolyethylene resembles petroleum-based polyethylene in structure and does not readily biodegrade. Conversely, some petroleum-derived polymers, such as polybutylene succinate, degrade under appropriate conditions [1].

This distinction matters. Mislabeling materials as environmentally benign without a full lifecycle evaluation can intensify pollution risks. Global bioplastic production continues to expand, yet non-biodegradable bio-based plastics account for a substantial fraction of output [1]. Production growth must therefore align with responsible end-of-life systems.

9.1.2 *Standards and the Limits of Certification*

International standards define compostability and biodegradability under controlled conditions. The European standard EN 13432 requires 90% conversion of carbon to carbon dioxide within 6 months under industrial composting conditions. ISO 17556 evaluates aerobic biodegradation in soil. EN 17033 governs biodegradable mulch films for agricultural use [1].

These standards provide measurable benchmarks. However, they simulate optimized environments with regulated temperature, humidity, and microbial activity. Field conditions differ widely. Soil temperature, moisture, microbial diversity, and management practices alter degradation rates [2]. A material certified as compostable may persist for years in marine systems. Polylactic acid bottles have shown half-lives exceeding several decades in aquatic environments [7]. Researchers now describe biodegradability as a system property that integrates polymer chemistry and environmental context [4]. This concept shifts evaluation from laboratory tests toward real-world performance.

9.1.3 Biodegradable Mulch Films in Agriculture

Agriculture illustrates both promise and complexity. Conventional low-density polyethylene mulches improve crop yields, conserve soil moisture, and suppress weeds. Yet they accumulate in soils and fragment into microplastics when removal remains incomplete [8].

Biodegradable mulch films based on starch blends, polybutylene adipate terephthalate, polybutylene succinate, polylactic acid, and polyhydroxyalkanoates aim to reduce residual plastic burdens [1]. After harvest, farmers till these films into the soil where microbial communities initiate degradation.

Evidence shows variable outcomes. Warmer temperatures accelerate biodegradation, while cooler climates slow the process [2]. Some polymers, such as PLA, degrade slowly in typical soil conditions because their glass transition temperatures exceed average field temperatures [9]. Residues of PBAT-based films have persisted in agricultural soils for multiple seasons [8].

Fragmentation into micro-bioplastics may occur before complete mineralization. Exposure to ultraviolet radiation enhances chain scission and particle release [5]. In marine deployment experiments, compostable bags degraded more rapidly than conventional plastics, yet many biodegradable materials remained structurally intact for extended periods [6]. These findings underscore that biodegradability does not eliminate microplastic risk. It modifies the degradation pathway and timescale.

9.1.4 Additives and Chemical Complexity

Polymer backbones rarely act alone. Manufacturers incorporate plasticizers, ultraviolet stabilizers, antioxidants, and other additives to improve flexibility, durability, and crop performance. Phthalates, benzophenones, and related compounds may migrate into soils and water [1].

Accelerated fragmentation of biodegradable polymers may enhance the release rates of additives compared with those of conventional plastics [10]. Leachates from both conventional and biodegradable microplastics have induced oxidative stress responses in aquatic organisms [11].

Thus, sustainable packaging and agricultural films must address polymer chemistry and additive profiles simultaneously. Green material design requires reducing hazardous additives and transparently disclosing the composition.

9.1.5 Soil Ecosystem Response

When biodegradable films are incorporated into soil, they interact with complex biogeochemical cycles. Studies report changes in soil aggregate stability, water retention, and nutrient dynamics following exposure to microbioplastic residues [8, 12].

Microbial communities respond to new carbon substrates and attachment surfaces. Shifts in bacterial taxa such as Proteobacteria and Actinobacteria have been observed in soils amended with biodegradable residues [13]. Enhanced enzyme activities may stimulate carbon cycling, yet long-term effects on soil organic matter remain uncertain [14].

These transformations illustrate that biodegradable plastics actively reshape soil ecology rather than disappear without a trace. Sustainable innovation must therefore evaluate agronomic benefits alongside ecological stability.

9.1.6 Circular Economy Pathways

Circular economy strategies extend beyond material substitution. They prioritize reduction, reuse, and redesign. Mechanical recycling, chemical depolymerization, and improved collection systems reduce environmental leakage.

For biodegradable materials, circularity requires compatibility with existing waste infrastructure. Industrial composting facilities must operate effectively and remain accessible. In regions lacking formal composting systems, biodegradable packaging may enter unmanaged landfills or open environments where degradation slows dramatically [15, 16].

Emerging economies face unique challenges. Limited waste segregation, informal recycling sectors, and inadequate infrastructure increase environmental leakage. Circular strategies must therefore integrate technological innovation with governance, investment, and community engagement.

9.1.7 Toward Truly Sustainable Packaging

Sustainable packaging design demands a systems approach. It integrates feedstock sourcing, polymer chemistry, additive safety, product performance, waste management compatibility, and ecological impact.

Evidence shows that biodegradable polymers, such as polyhydroxyalkanoates, mineralize more readily in soil than some alternatives [17]. Yet mineralization rates vary widely across environments. Transparent lifecycle assessments and long-term field experiments remain essential.

Innovation succeeds when it aligns chemistry with ecology. Bioplastics and sustainable packaging can reduce persistent plastic accumulation. However, they require rigorous standards, additive reformulation, real-world testing, and integration into circular waste systems.

Technological optimism must therefore rest on scientific evidence. When design anticipates environmental fate and infrastructure constraints, bioplastics and circular economy strategies can shift plastic management from reactive cleanup to preventive sustainability.

9.2 Waste Management and Recycling Transitions in Emerging Economies

Waste management systems in emerging economies stand at a decisive turning point. Rapid urbanization, expanding consumer markets, and limited infrastructure have intensified plastic pollution in rivers, coastal waters, and agricultural landscapes. These conditions accelerate the generation of microplastics through fragmentation, open burning, and uncontrolled disposal. The transition toward sustainable waste management is therefore central to mitigating microplastic pollution and protecting environmental and human health.

9.2.1 Structural Gaps in Waste Systems

Many emerging economies rely on landfilling and open dumping as dominant disposal pathways. Inadequate collection coverage and limited segregation at source permit plastics to enter waterways, where mechanical abrasion, ultraviolet exposure, and microbial activity drive their breakdown into microplastics [18]. Wastewater treatment plants remove a large fraction of microplastics, yet even a small discharge percentage translates into billions of particles released annually due to high influent volumes [19]. These findings demonstrate that conventional infrastructure cannot fully prevent the dissemination of microplastics.

Plastic mulch films in agriculture further complicate waste streams. Conventional polyethylene films often persist in soils after harvest, fragmenting over time and contributing to the formation of secondary microplastics [1]. Residual fragments alter soil structure and influence aggregate formation, potentially affecting water retention and nutrient cycling. The persistence of these residues highlights the need for integrated waste strategies that extend beyond urban settings into agroecosystems.

9.2.2 Recycling Limitations and Informal Economies

Mechanical recycling in emerging economies faces technical and economic barriers. Contamination by organic matter, additives, and mixed polymer streams reduces recyclate quality. Films used in agriculture and food packaging are particularly difficult to recycle because soil residues and degradation products limit material recovery [1]. When recycling fails, plastics are diverted to incineration or open burning, processes that fragment materials and release microplastics into the air and ash residues.

Informal waste sectors play a substantial role in plastic recovery across Southeast Asia, Africa, and Latin America. Waste pickers recover high-value polymers such as polyethylene terephthalate and polypropylene, yet low-value multilayer plastics often escape collection. These unmanaged fractions contribute disproportionately to environmental leakage. Strengthening inclusive recycling frameworks, with formal recognition and occupational safeguards, can reduce leakage while enhancing social equity.

9.2.3 Transition Toward Circular Economy Models

Circular economy frameworks offer a strategic pathway to reduce microplastic generation. By redesigning products for durability, reuse, and recyclability, material loops can close before plastics enter the environment. Standardization of polymer types and reduction of hazardous additives improve recyclability and reduce toxic leachates [18].

Advanced sorting technologies, including spectroscopic and hyperspectral imaging systems, enhance polymer identification and separation efficiency [18]. Integration of such analytical tools into recycling facilities can minimize cross-contamination and improve the purity of recyclate. Although capital-intensive, regional material recovery facilities equipped with optical sorting represent a transformative investment for emerging economies seeking to modernize recycling infrastructure.

9.2.4 Nanotechnology and Innovative Removal Strategies

Emerging research highlights the potential of nanomaterials for microplastic removal from wastewater. Functionalized graphene, magnetic nanoparticles, and metal-organic frameworks demonstrate high adsorption efficiencies under laboratory conditions [19]. Magnetic nanocomposites enable rapid separation of bound microplastics, offering a scalable concept for tertiary treatment stages.

Three-dimensional reduced graphene oxide exhibits strong π–π interactions with polystyrene microplastics, facilitating adsorption in aquatic matrices [19]. Similarly, photocatalytic nanomaterials promote degradation of polymer surfaces under light irradiation, potentially transforming microplastics into smaller, mineralizable fragments [18]. While promising, these technologies require further validation under real wastewater conditions and cost-benefit assessment before large-scale deployment.

9.2.5 Agricultural Transitions and Biodegradable Alternatives

Biodegradable mulch films have emerged as alternatives to polyethylene films in agriculture. Standards such as EN 17033 specify criteria for biodegradation in soil environments [1]. These materials, composed of polymers such as polylactic acid, polybutylene adipate terephthalate, and starch blends, aim to reduce residual plastic accumulation.

However, incomplete degradation may release micro-bioplastics and additives into soils [1]. Field studies indicate that degradation rates depend on soil temperature, microbial diversity, and climatic conditions. Therefore, transitions to biodegradable plastics must be coupled with long-term field monitoring and clear regulatory oversight. Waste management strategies must account for both conventional plastics and emerging bioplastics to prevent unintended environmental burdens.

9.2.6 Policy Integration and Regional Cooperation

Effective waste transitions demand policy coherence across sectors. Extended producer responsibility schemes incentivize design for recyclability and fund collection systems. Deposit-return systems increase recovery rates of beverage containers, reducing leakage into aquatic systems. Harmonized standards for compostability and biodegradability clarify end-of-life pathways and prevent misleading labeling.

Regional cooperation within Southeast Asia and other emerging regions can strengthen monitoring networks and share technological expertise. Standardized detection protocols improve comparability of microplastic data, supporting evidence-based policy decisions [18]. Cross-border initiatives addressing riverine plastic transport are essential because major rivers serve as conduits from inland waste sources to coastal ecosystems.

9.2.7 Socioeconomic Dimensions of Transitions

Waste transitions intersect with poverty alleviation, public health, and employment. Upgrading recycling systems generates green jobs and reduces occupational hazards associated with informal waste picking. Public education campaigns promoting segregation at source enhance recycling efficiency and reduce contamination. Community-based monitoring initiatives empower citizens to track plastic leakage and advocate for infrastructure improvements. Investments in waste management infrastructure yield long-term economic benefits by protecting fisheries, tourism, and agriculture from microplastic contamination. In coastal nations, safeguarding marine ecosystems preserves livelihoods dependent on aquaculture and capture fisheries. Integrated waste governance, therefore, contributes to the Sustainable Development Goals related to health, clean water, and life below water. To synthesize these interconnected dimensions, Fig. 9.1 presents an integrated framework of waste management and recycling transitions in emerging economies.

The figure illustrates how structural gaps, recycling limitations, agricultural plastic use, technological innovations, policy coordination, and socioeconomic drivers interact to shape the transition from linear waste systems toward circular resource

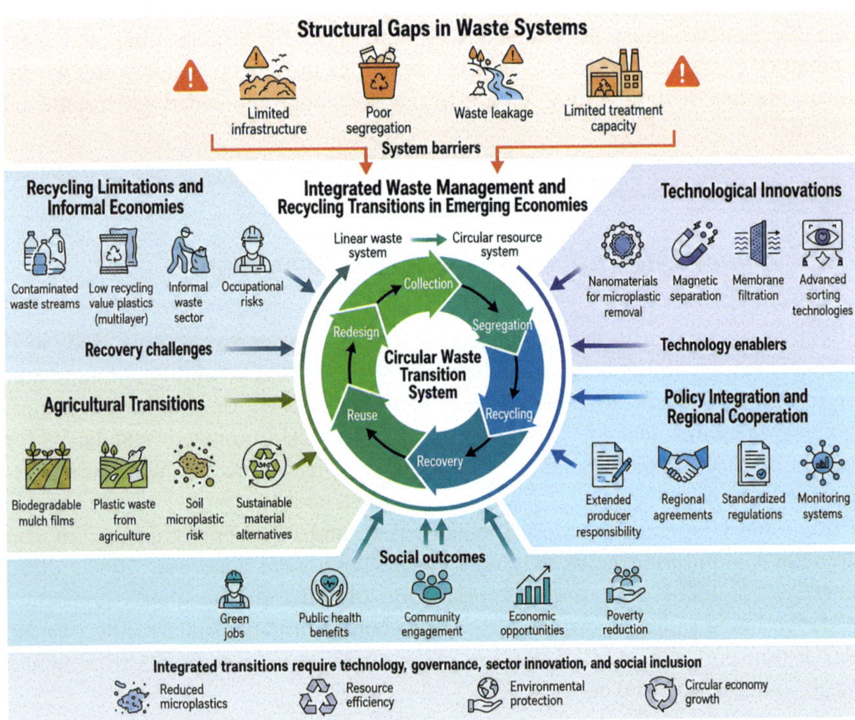

Fig. 9.1 Integrated waste management and recycling transition framework in emerging economies. (Conceptual framework developed by the authors and illustrated using digital design tools)

management. It highlights that effective microplastic mitigation requires coordinated advances in infrastructure, governance, technological innovation, and social inclusion rather than isolated interventions.

9.2.8 Path Forward

Emerging economies possess the opportunity to leapfrog linear waste models and adopt integrated circular systems. Combining improved collection, advanced recycling technologies, nanomaterial-assisted removal strategies, and responsible agricultural practices can substantially reduce microplastic emissions. Scientific innovation must align with policy reform and community engagement to achieve measurable impact.

The transition toward sustainable waste management represents a cornerstone of microplastic mitigation. By embedding scientific advances within inclusive governance frameworks, emerging economies can transform waste from an environmental liability into a resource stream that supports resilient, plastic-conscious futures.

9.3 Advanced Cleanup Technologies: From Ocean Skimmers to Nanotechnology

Cleanup technologies sit at a difficult intersection. They must remove plastics from water without harming ecosystems, while recognizing that plastics move across five major marine compartments: coastlines, surface waters, the water column, the seabed, and biota [20]. Currents and winds can concentrate floating debris in convergence zones far from where it entered, and buoyant items can later sink as biofouling changes density and drag [20]. These dynamics explain a core engineering challenge: capture systems must match where plastics actually travel, not where people imagine them. Effective innovation, therefore, follows a simple logic: intercept early, separate efficiently, and treat residues responsibly.

9.3.1 Ocean Skimmers and Surface Capture in Coastal Waters

Mechanical recovery still matters, especially for larger debris that later fragments into smaller particles. In sheltered settings, such as marinas, ports, and coastal embayments, floating skimmer units can remove buoyant litter before wave action and sunlight accelerate fragmentation. The attached literature highlights the Seabin concept as a floating waste collector and pairs it with "intercept systems" designed to capture plastics before they degrade into microplastics

(MPs) [20, 21]. These systems succeed where conditions remain controlled: low wave energy, predictable circulation, and easy access for maintenance. Operators can then treat captured material as a managed waste stream, rather than a dispersed pollutant.

Yet surface skimmers face two hard limits. First, they target the buoyant fraction, while many polymers and biofouled particles occupy the water column or seabed [20]. Second, skimmers do not directly remove microplastics that are already dispersed into millimeter- and micrometer-sized particles. For these fractions, engineering shifts from "collect by size" to "collect by surface chemistry," which explains the rise of advanced separations and nano-enabled sorbents.

9.3.2 Interception Before the Ocean: The Highest Leverage "Cleanup"

The most cost-effective cleanup often occurs upstream. Plastics continuously enter waters through domestic litter and riverine pathways, and extreme events such as floods and storms can deliver large pulses of waste [20]. In practice, interception technologies aim to convert diffuse flow into a capture point: booms, screens, and guided flows that concentrate floating debris for removal. This approach aligns with the evidence base that chronic land-based sources provide near continuous inputs to the sea [22].

A key advantage of interception lies in its operational practicality. Rivers and drainage canals act as defined transport pathways where capture technologies can be strategically positioned to facilitate routine collection, maintenance, and downstream recovery operations. The inclusion of Ocean Cleanup Interception B.V. among the funding partners further highlights the increasing prioritization of interception technologies and engineered recovery approaches as critical interventions to prevent plastic leakage into marine environments [23].

9.3.3 Separation Technologies That "Grab" Microplastics

When plastics shrink to the micro-scale, successful removal depends on how particles interact with water treatment steps. Evidence from water and wastewater studies shows that conventional operations can remove large fractions of microplastics, but performance varies with polymer type, size, and surface weathering [24, 25]. This variability explains why advanced cleanup now focuses on tunable capture mechanisms.

9.3.3.1 Coagulation and Flocculation as Engineered Capture

Coagulation can destabilize particles and drive aggregation into settleable flocs. In municipal contexts, studies report that removal efficiencies depend on particle weathering and the choice of coagulant, with polymer surfaces influencing attachment and settling behavior [26]. Process optimization, therefore, becomes a microplastics-specific design problem rather than a simple reuse of legacy turbidity control.

9.3.3.2 Membrane-Based Recovery as a Barrier Approach

Membrane filtration provides a physical cutoff and offers an attractive option for centralized systems, especially where reuse and high effluent standards are important. Reviews in environmental water technology describe membrane remediation pathways for microplastics and clarify why fouling control and concentrate management determine feasibility [27].

9.3.3.3 Magnetic Extraction as a Rapid "Switch"

Magnetic recovery stands out because it converts separation into a simple field-driven response. A clear proof of concept is provided by studies demonstrating the magnetic extraction of microplastics from environmental samples [28]. In practice, magnetic methods often rely on functionalized materials that selectively bind to plastic surfaces, enabling the resulting composites to be removed using an external magnetic field. This approach reduces reliance on high-pressure filtration, minimizes clogging issues, and can significantly accelerate sample processing, particularly when dealing with complex environmental matrices containing mixed organic and inorganic debris.

9.3.4 Nanotechnology: Turning Microplastics into a Controllable Target

Nano-enabled cleanup aims to solve what skimmers cannot: the dispersed, small, and chemically diverse microplastic fraction. The attached sources describe a fast-growing toolkit: engineered sorbents, surface-active composites, and hybrid systems that combine capture with degradation or regeneration [29]. Critically, nanotechnology does not mean "smaller filters." It means engineered interfaces that recognize and bind microplastics through hydrophobic interactions, electrostatics, roughness matching, and tailored porosity.

9.3.4.1 Adsorptive Nanomaterials and Structured Sorbents

Researchers increasingly design porous solids that trap plastics while allowing water to pass. Examples in the attached literature include materials chemistry approaches for adsorptive capture and selective interactions with microplastic-like particles [30]. Such systems can work as packed beds, coated meshes, or composite foams. Engineers then evaluate regeneration, pressure drop, and sorbent longevity as practical constraints rather than laboratory curiosities.

9.3.4.2 Nano-Enabled Hybrid Remediation

Many promising concepts combine capture, concentration, and treatment. Coagulation can pre-concentrate particles, membranes can polish effluent, and sorbents can reduce the remaining microplastic load. The most realistic pathway is system integration, in which each unit operation reduces the burden on the next. This layered strategy mirrors how modern drinking water and wastewater plants already manage complex contaminants [24, 25].

9.3.5 Designing the "Cleanup Stack" for the Next Decade

Advanced cleanup now looks less like a single device and more like a coordinated stack:

1. *Intercept and skim* at river mouths, canals, and coastal hotspots to prevent fragmentation and reduce ocean loading [20, 23].
2. *Engineer separation* in treatment systems using coagulation, membrane barriers, and magnetic extraction to capture diverse particle classes [26–28].
3. *Deploy nano-enabled materials* where small particle recovery becomes the limiting step, while enforcing safety and lifecycle governance [29].
4. *Close the loop* by treating captured plastics as managed feedstocks for valorization and responsible disposal, rather than returning them to the environment through poorly controlled handling.

This integrated approach recognizes a truth often ignored in public debates: the ocean is not a contained reactor. Plastics circulate, sink, resurface, and fragment across compartments. Cleanup succeeds when it matches that movement with smart interception, targeted separations, and materials innovation guided by safety and systems engineering [20, 29].

9.4 Mitigation Strategies for Agriculture and Aquaculture Sectors

Microplastics control in agriculture and aquaculture is most effective when prevention and removal work together. These sectors sit at a critical junction: they rely on plastics for efficiency, yet they operate in open systems where fragmentation, runoff, and gear wear turn convenience into chronic contamination. Effective mitigation, therefore, targets the full pathway: inputs (what enters the farm), transformation (how plastics break down into smaller pieces), and export (how particles move into waterways and into seafood). Evidence from recent syntheses and field studies shows that the highest returns come from source control, tighter material standards, and practical capture technologies integrated into routine operations [31–33].

9.4.1 *Agriculture: Stop Microplastics at the Farm Gate*

1. *Redesign Plastic Mulch Use, Retrieval, and Accountability*
   Plastic mulch films remain a dominant agricultural source of microplastics because residues persist after harvest, especially when films tear, thin, or embed into soil. A strong mitigation package starts with material selection and ends with verified retrieval. Thicker and more tear-resistant films reduce fragmentation; standardized removal protocols and retrieval audits reduce field residues; and take-back schemes create incentives to recover film rather than abandon it [31, 34]. The goal is simple: treat mulch as managed equipment, not as disposable input.
2. *Use of Biodegradable Mulches with Stricter Performance and Residue Criteria*
   Biodegradable mulches can lower visible plastic accumulation, yet they do not automatically eliminate microplastic risk. Studies show that biodegradable mulch residues can still contribute to microplastics in soil, with measurable effects on crop growth indicators and soil properties under some exposure conditions [35]. Mitigation, therefore, needs procurement standards that specify field degradation behavior, residue thresholds, and end-of-season verification. Farmers and regulators should require performance labeling tied to local climate and soil conditions, supported by monitoring plans that verify residue decline rather than assume it [35, 36].
3. *Control Biosolids and Compost Pathways with Treatment, Screening, and Alternative Routing*
   Wastewater treatment captures a large fraction of incoming microplastics in sludge, which can then transfer to agricultural soils when biosolids are land-applied [32]. Field evidence supports long residence times in soils and very low vertical infiltration rates over short observation windows, suggesting microplastics can accumulate rather than quickly disappear [32]. Mitigation should therefore include a "clean biosolids" program: upstream source controls, in-plant

capture, and post-treatment separation before land application. A recent review identifies interventions across the chain, including microfiber filtration devices and optimized separation processes, and highlights promising options, such as froth flotation, high gradient magnetic separation, pyrolysis, and related coupled approaches to reduce microplastics in biosolids prior to soil application [37].

4. *Strengthen Farm Water Barriers to Reduce Irrigation and Runoff Transport*
   Microplastics can enter soils through irrigation water and leave farms through runoff and drainage. Since runoff can mobilize a fraction of soil microplastics, even when export rates appear modest, catchment-scale control remains important [32]. Practical barriers include sediment traps, vegetated buffer strips, and engineered filtration for recycled irrigation water in high-risk settings. Where feasible, integrating fine filtration steps into water reuse loops provides dual benefits, reducing both microplastics and suspended solids that carry contaminants.
5. *Shift Packaging and On-Farm Plastics Toward Circular Designs with Verified End-of-Life*
   Agriculture uses a wide range of plastics beyond mulch: seedling trays, fertilizer sacks, silage wrap, irrigation components, and packaging. Mitigation improves when procurement favors reusable systems, high-quality polymers with stable performance, and traceable collection. Reviews on bioplastics and sustainable packaging emphasize that environmental benefits depend on realistic infrastructure, labeling, and end-of-life routes, rather than polymer choice alone [36]. This points to a practical policy direction: require suppliers to provide retrieval logistics and verified recycling or treatment pathways as part of agricultural plastic sales.

9.4.2 Aquaculture: Reduce Gear-Derived Particles and Treat Water Like a Product Quality

Aquaculture faces a different risk profile than terrestrial farms. Microplastics arrive from ambient waters, but they also originate within facilities through ropes, nets, floats, liners, and packaging. A large review of microplastics in aquaculture highlights both external inputs and aquaculture process sources, and it frames control around ecological interception, purification, and improved operational management [33].

1. *Cut Primary Shedding from Gear Through Design, Maintenance, and Procurement*
   Net abrasion and rope wear generate fragments and fibers, especially in high-energy systems. Mitigation starts with gear standards: abrasion-resistant materials, protective coatings where appropriate, and maintenance schedules that prevent brittle failure. Procurement should favor longer-life gear with documented wear performance, coupled with gear-tracking and recovery protocols

that reduce loss during storms and routine operations. These steps reduce both microplastics and macroplastic debris that later fragments.

2. *Close the Loop Through Recycling and Responsible End-of-Life for Nets and Ropes*

   Gear take-back programs and regional recycling hubs reduce incentives to discard damaged nets. Facilities can implement simple inventory systems that link gear issuance to return. These are operational controls, yet they produce measurable prevention outcomes because lost gear continues shedding for years.

3. *Integrate Purification Into Production Systems, Especially in Recirculating Aquaculture*

   Recirculating aquaculture systems provide an engineering advantage: water passes through controlled treatment units where particles can be removed. Membrane-based RAS approaches use ultrafiltration and membrane bioreactors to manage water quality, enabling the removal of fine particulates while reducing water exchange [38]. Coupling filtration with optimized biofilters can simplify setups and stabilize water quality, supporting intensified farming with lower discharge [39]. While these studies focus on overall water treatment efficiency, the same unit operations can reduce particulate loads, including microplastics, especially when combined with targeted capture steps [38, 39].

4. *Treat Wastewater and Effluents with High-Performance Particle Capture*

   Where ponds or flow-through systems discharge water, add separation steps at outlet points. Electrocoagulation has strong evidence as a microplastics removal technology in aquatic settings and can complement filtration in waste streams [40, 41]. These technologies become most feasible when deployed at centralized discharge points, such as shared treatment facilities in aquaculture clusters.

5. *Use Ecological Interception as a Proactive Barrier in Coastal Aquaculture Landscapes*

   For coastal farms, natural buffers can reduce the transport of microplastics to open waters. Mangrove systems can function as interception zones where microplastics accumulate in sediments through repeated suspension and reprecipitation dynamics [42]. This supports a nature-based mitigation pathway: maintain and restore mangroves near aquaculture areas to reduce outward transport of plastics while improving shoreline stability and water quality co-benefits [42].

6. *Align Food Safety Controls with Microplastics Monitoring*

   Aquaculture products face increasing scrutiny. The aquaculture review literature emphasizes monitoring and standardization as future priorities for effective control [33]. Farms can embed microplastics risk into quality assurance by selecting low-shedding materials, controlling packaging fibers, and using periodic screening in water, sediments, and representative organisms in high-risk sites.

9.4.3 A Practical Mitigation Ladder for Farms

Across agriculture and aquaculture, mitigation scales best when it follows a clear ladder: avoid, substitute, capture, and verify. Avoid practices that generate residues, such as thin films without retrieval. Substitute only when performance and end-of-life are verified, especially for biodegradable materials in soils. Capture microplastics at controllable nodes, including biosolids treatment, effluent filtration, and RAS purification. Verify through monitoring plans tied to procurement and operations. When farms adopt this ladder, microplastics control becomes part of productivity and quality, rather than an added burden.

9.5 Chapter Synthesis

This chapter highlights how scientific progress and technological innovation are reshaping the response to microplastic pollution. Advances in sustainable material science, circular economy models, and improved waste management systems demonstrate that prevention remains the most effective long-term strategy. Developments in biodegradable polymers, alternative packaging materials, and product redesign show how upstream interventions can reduce environmental leakage before pollution occurs. These approaches also emphasize the need to redesign production systems so that environmental protection becomes an integral component of industrial development rather than an afterthought.

Progress in waste management transitions further illustrates how technological improvements can reduce plastic losses across the value chain. Expanded recycling systems, material recovery facilities, and improved segregation practices show strong potential to reduce secondary microplastic formation. Emerging economies face structural challenges, including infrastructure gaps, rapid urbanization, and limited financial resources. Yet these same regions also offer opportunities for leapfrogging innovations through decentralized recycling systems, community-based recovery programs, and digital monitoring platforms that improve collection efficiency. These developments reinforce the importance of aligning technological solutions with socioeconomic realities.

Engineering innovations also play a critical role in remediation. Advances in filtration systems, adsorption technologies, electrostatic separation methods, and nanotechnology-based capture materials provide new tools for removing microplastics from water, wastewater, and industrial discharges. Ocean cleanup devices, river interception systems, and stormwater filtration technologies demonstrate how engineering solutions can intercept plastics before they fragment and become more environmentally persistent. While these solutions cannot replace prevention strategies, they provide essential complementary tools for reducing existing pollution burdens.

Sector-specific mitigation strategies further demonstrate the importance of targeted interventions. Improvements in agricultural plastics management, biodegradable mulch films, and better recovery of aquaculture materials show how pollution prevention can be integrated into food production systems. These approaches highlight the need for sector-driven solutions that consider operational realities while reducing environmental impacts. Scientific collaboration across materials science, environmental engineering, and industrial ecology continues to expand the range of feasible mitigation pathways.

Technological progress, therefore, reflects a broader transition toward solution-oriented environmental science. Innovation increasingly connects research laboratories with real-world implementation through pilot systems, demonstration projects, and industry partnerships. This transition also emphasizes scalability, affordability, and adaptability, particularly in regions where resource constraints require practical, cost-effective solutions. The growing convergence of science, engineering, and sustainability policy demonstrates that technological responses can support systemic environmental change when supported by appropriate governance frameworks and market incentives.

These advances establish an important foundation for the next phase of the discussion. Technical solutions and policy frameworks create the conditions for change, yet long-term success depends on how societies adopt and sustain these interventions. The next chapter, therefore, shifts attention toward the role of communities, individual behavior, and collective environmental responsibility. It explores how grassroots initiatives, citizen science, education, and lifestyle choices can translate scientific knowledge and technological progress into sustained environmental action.

References

1. Campanale, C., Galafassi, S., Di Pippo, F., Pojar, I., Massarelli, C., & Uricchio, V. F. (2024). A critical review of biodegradable plastic mulch films in agriculture. *Trends in Analytical Chemistry, 170*, 117391. https://doi.org/10.1016/j.trac.2023.117391
2. Sintim, H. Y., Bary, A. I., Hayes, D. G., Wadsworth, L. C., Anunciado, M. B., English, M. E., Bandopadhyay, S., Schaeffer, S. M., DeBruyn, J. M., & Flury, M. (2020). In situ degradation of biodegradable plastic mulch films in agricultural soils. *Science of the Total Environment, 727*, 138668. https://doi.org/10.1016/j.scitotenv.2020.138668
3. Solanki, P., Jain, S., Mehrotra, R., Mago, P., & Dagar, S. (2024). Microplastics in agricultural soil and their impact: A review. *Nature Environment and Pollution Technology, 23*(4), 2143–2155. https://doi.org/10.46488/NEPT.2024.v23i04.019
4. Science Advice for Policy by European Academies (SAPEA). (2020). *Biodegradability of plastics in the open environment*. SAPEA. https://doi.org/10.26356/biodegradabilityplastics
5. Lambert, S., & Wagner, M. (2017). Environmental performance of bio-based and biodegradable plastics: The road ahead. *Chemical Society Reviews, 46*(22), 6855–6871. https://doi.org/10.1039/C7CS00149E
6. Napper, I. E., & Thompson, R. C. (2019). Environmental deterioration of biodegradable, oxo-biodegradable, compostable, and conventional plastic carrier bags in the sea, soil, and

open-air over a 3-year period. *Environmental Science & Technology, 53*(9), 4775–4783. https://doi.org/10.1021/acs.est.8b06984
7. Ali, W., Ali, H., Gillani, S., Zinck, P., & Souissi, S. (2023). Polylactic acid synthesis, biodegradability, conversion to microplastics and toxicity: A review. *Environmental Chemistry Letters, 21*, 1761–1786. https://doi.org/10.1007/s10311-022-01562-9
8. Qiang, L., Hu, H., Li, G., Xu, J., Cheng, J., Wang, J., & Zhang, R. (2023). Plastic mulching, and occurrence, incorporation, degradation, and impacts of polyethylene microplastics in agroecosystems. *Ecotoxicology and Environmental Safety, 263*, 115274. https://doi.org/10.1016/j.ecoenv.2023.115274
9. Liu, E., Zhang, L., Dong, W., & Yan, C. (2021). Biodegradable plastic mulch films in agriculture: Feasibility and challenges. *Environmental Research Letters, 16*(1), 011004. https://doi.org/10.1088/1748-9326/abd211
10. Zimmermann, L., Dombrowski, A., Völker, C., & Wagner, M. (2020). Are bioplastics and plant-based materials safer than conventional plastics? In vitro toxicity and chemical composition. *Environment International, 145*, 106066. https://doi.org/10.1016/j.envint.2020.106066
11. González-Pleite, M., Tamayo-Belda, M., Pulido-Reyes, G., Amariei, G., Leganés, F., Rosal, R., & Fernández-Piñas, F. (2019). Secondary nanoplastics released from a biodegradable microplastic severely impact freshwater environments. *Environmental Science: Nano, 6*(5), 1382–1392. https://doi.org/10.1039/C8EN01427B
12. Chae, Y., & An, Y.-J. (2018). Current research trends on plastic pollution and ecological impacts on the soil ecosystem: A review. *Environmental Pollution, 240*, 387–395. https://doi.org/10.1016/j.envpol.2018.05.008
13. Yang, C., Huang, Y., Long, B., & Gao, X. (2022). Effects of biodegradable and polyethylene film mulches and their residues on soil bacterial communities. *Environmental Science and Pollution Research, 29*, 89698–89711. https://doi.org/10.1007/s11356-022-21334-9
14. Zhou, Y., Wang, J., Zou, M., Jia, Z., Zhou, S., & Li, Y. (2020). Microplastics in soils: A review of methods, occurrence, fate, transport, ecological and environmental risks. *Science of the Total Environment, 748*, 141368. https://doi.org/10.1016/j.scitotenv.2020.141368
15. Mosconi, E. M., Colantoni, A., & Tarantino, M. (2024). Strategies for enhancing the efficiency of packaging and managing packaging waste in compliance with regulations. In *IntechOpen eBooks*. IntechOpen. https://doi.org/10.5772/intechopen.1007064
16. Phillip, A. (2024). Bioplastics from waste biomass: Paving the way for a sustainable future. *International Journal for Research in Applied Science and Engineering Technology, 12*(9), 518. https://doi.org/10.22214/ijraset.2024.64225
17. Clarke, R. W., Rosetto, G., Uekert, T., Curley, J. B., Moon, H., Knott, B. C., McGeehan, J. E., & Knauer, K. M. (2024). Polyhydroxyalkanoates in emerging recycling technologies for a circular materials economy. *Materials Advances, 5*(17), 6690. https://doi.org/10.1039/d4ma00411f
18. Li, M., Zhao, Z., Zhao, Z., & Li, M. (2025). Review of techniques for the detection, removal, and transformation of environmental microplastics and nanoplastics. *ACS Applied Materials & Interfaces, 17*(14), 20560–20589. https://doi.org/10.1021/acsami.5c02306
19. Indhur, R., Amoah, I., Bux, F., & Kumari, S. (2023). Nanomaterials for microplastic removal from wastewater: Current state of the art nanomaterials and future prospects. *ACS ES&T Water, 3*(12), 3741–3754. https://doi.org/10.1021/acsestwater.3c00203
20. Van Sebille, E., Wilcox, C., Lebreton, L., Maximenko, N., Hardesty, B. D., van Franeker, J. A., Eriksen, M., Siegel, D., Galgani, F., & Law, K. L. (2015). A global inventory of small floating plastic debris. *Environmental Research Letters, 10*(12), 124006. https://doi.org/10.1088/1748-9326/10/12/124006
21. Rochman, C. M., Brookson, C., Bikker, J., Djuric, N., Earn, A., Bucci, K., Athey, S., Huntington, A., McIlwraith, H., & Munno, K. (2019). Rethinking microplastics as a diverse contaminant suite. *Environmental Toxicology and Chemistry, 38*(4), 703–711. https://doi.org/10.1002/etc.4371

22. Siegfried, M., Koelmans, A. A., Besseling, E., & Kroeze, C. (2017). Export of microplastics from land to sea. A modelling approach. *Water Research, 127*, 249–257. https://doi.org/10.1016/j.watres.2017.10.011
23. Tan, H., Othman, M. H. D., Chong, W. T., Kek, H. Y., Wong, S. L., Nyakuma, B. B., Mong, G. R., Abdul Wahab, R., & Wong, K. Y. (2024). Turning plastics/microplastics into valuable resources? Current and potential research for future applications. *Journal of Environmental Management, 356*, 120644. https://doi.org/10.1016/j.jenvman.2024.120644
24. Conley, K., Clum, A., Deepe, J., Lane, H., & Beckingham, B. (2019). Wastewater treatment plants as a source of microplastics to an urban estuary: Removal efficiencies and loading per capita over one year. *Water Research X, 3*, 100030. https://doi.org/10.1016/j.wroa.2019.100030
25. Mason, S. A., Garneau, D., Sutton, R., Chu, Y., Ehmann, K., Barnes, J., Fink, P., Papazissimos, D., & Rogers, D. L. (2016). Microplastic pollution is widely detected in US municipal wastewater treatment plant effluent. *Environmental Pollution, 218*, 1045–1054. https://doi.org/10.1016/j.envpol.2016.08.056
26. Lapointe, M., Farner, J. M., Hernandez, L. M., & Tufenkji, N. (2020). Understanding and improving microplastic removal during water treatment: Impact of coagulation and flocculation. *Environmental Science & Technology, 54*(14), 8719–8727. https://doi.org/10.1021/acs.est.0c00712
27. Malankowska, M., Echaide-Gorriz, C., & Coronas, J. (2021). Microplastics in marine environment: A review on sources, classification, and potential remediation by membrane technology. *Environmental Science: Water Research & Technology, 7*, 243–258. https://doi.org/10.1039/D0EW00802H
28. Grbić, J., Nguyen, B., Guo, E., You, J. B., Sinton, D., & Rochman, C. M. (2019). Magnetic extraction of microplastics from environmental samples. *Environmental Science & Technology Letters, 6*(2), 68–72. https://doi.org/10.1021/acs.estlett.8b00671
29. Sanjeev, N. O., Vallabha, M. S., & Rabi, R. R. L. (2025). Nanotechnology-based approaches for the removal of microplastics from wastewater: A comprehensive review. *Beilstein Journal of Nanotechnology, 16*, 1607–1632. https://doi.org/10.3762/bjnano.16.114
30. Chen, Y.-J., Chen, Y., Miao, C., Wang, Y.-R., Gao, G.-K., Yang, R.-X., Zhu, H.-J., Wang, J.-H., Li, S.-L., & Lan, Y.-Q. (2020). Metal–organic framework-based foams for efficient microplastics removal. *Journal of Materials Chemistry A, 8*, 14644–14652. https://doi.org/10.1039/D0TA04891G
31. Radford, F., Horton, A., Hudson, M., Shaw, P., & Williams, I. (2023). Agricultural soils and microplastics: Are biosolids the problem? *Frontiers in Soil Science, 2*, 941837. https://doi.org/10.3389/fsoil.2022.941837
32. Schell, T., Hurley, R., Buenaventura, N. T., Mauri, P. V., Nizzetto, L., Rico, A., & Vighi, M. (2022). Fate of microplastics in agricultural soils amended with sewage sludge: Is surface water runoff a relevant environmental pathway? *Environmental Pollution, 293*, 118520. https://doi.org/10.1016/j.envpol.2021.118520
33. Wu, H., Hou, J., & Wang, X. (2023). A review of microplastic pollution in aquaculture: Sources, effects, removal strategies and prospects. *Ecotoxicology and Environmental Safety, 252*, 114567. https://doi.org/10.1016/j.ecoenv.2023.114567
34. Sintim, H. Y., & Flury, M. (2017). Is biodegradable plastic mulch the solution to agriculture's plastic problem? *Environmental Science & Technology, 51*(3), 1068–1069. https://doi.org/10.1021/acs.est.6b06042
35. Liang, R., Zhang, C., Zhang, R., Li, Q., Liu, H., & Wang, X.-X. (2024). Effects of microplastics derived from biodegradable mulch film on different plant species growth and soil properties. *Science of the Total Environment, 948*, 174899. https://doi.org/10.1016/j.scitotenv.2024.174899
36. Kumar, S., Dubey, N., Kumar, V., Choi, I., Jeon, J., & Kim, M. (2024). Combating micro/nano plastic pollution with bioplastic: Sustainable food packaging, challenges, and future perspectives. *Environmental Pollution, 363*(Part 1), 125077. https://doi.org/10.1016/j.envpol.2024.125077
37. Bawa, S. A., Chan, A., Wrobel-Tobiszewska, A., Hardie, M., & Towns, C. (2024). A review of methods for mitigating microplastic contamination in biosolids from wastewater treatment

plants before agricultural soil application. *Science of the Total Environment, 957*, 177360. https://doi.org/10.1016/j.scitotenv.2024.177360
38. Widiasa, I. N., Susanto, H., Ting, Y. P., Suantika, G., Steven, S., Khoiruddin, K., & Wenten, I. G. (2024). Membrane-based recirculating aquaculture system: Opportunities and challenges in shrimp farming. *Aquaculture, 579*, 740224. https://doi.org/10.1016/j.aquaculture.2023.740224
39. Liu, W., Du, X., Tan, H., Xie, J., Luo, G., & Sun, D. (2021). Performance of a recirculating aquaculture system using biofloc biofilters with convertible water-treatment efficiencies. *Science of the Total Environment, 754*, 141918. https://doi.org/10.1016/j.scitotenv.2020.141918
40. Liu, F., Zhang, C., Li, H., Offiong, N.-A. O., Bi, Y., Zhou, R., & Ren, H. (2023). A systematic review of electrocoagulation technology applied for microplastics removal in aquatic environment. *Chemical Engineering Journal, 456*, 141078. https://doi.org/10.1016/j.cej.2022.141078
41. Perren, W., Wojtasik, A., & Cai, Q. (2018). Removal of microbeads from wastewater using electrocoagulation. *ACS Omega, 3*(3), 3357–3364. https://doi.org/10.1021/acsomega.7b02037
42. Liu, X., Liu, H., Chen, L., & Wang, X. (2022). Ecological interception effect of mangroves on microplastics. *Journal of Hazardous Materials, 423*(Part B), 127231. https://doi.org/10.1016/j.jhazmat.2021.127231

Chapter 10
Community and Individual Action

Introduction
Communities already hold the missing piece in microplastics control: eyes on the ground, local knowledge of hotspots, and the social power to shift norms. Citizen science and community monitoring can generate large, policy-relevant datasets across coastlines, rivers, and cities, yet projects often struggle with inconsistent methods, uneven participation, and weak translation of results into decisions [1, 2]. When programs adopt clear protocols, training, and verification, volunteers can map microplastic and litter patterns at scales that conventional teams rarely reach [3, 4].

Community action also works as prevention. Grassroots groups use monitoring results to push upstream fixes: pellet loss controls, local ordinances, stronger producer responsibility, and better waste services [4, 5]. Education-driven monitoring can shift perceptions and behaviors, especially among youth, when activities connect field observations to solutions and civic engagement [6].

10.1 Grassroots Movement and Community-Based Monitoring

Grassroots microplastics action begins with a simple truth: plastic pollution shows up first in neighborhoods, shorelines, fishing grounds, and rivers where people live. Local groups often detect problems before formal monitoring does, because they walk the same transects weekly, notice changes after storms, and track the sources that outsiders miss. Citizen science, therefore, fits microplastics work as both a data engine and a governance tool. Systematic reviews stress that citizen science can cover "source to sea" pathways, yet data quality, methodological inconsistency, and limited policy uptake frequently constrain impact [1].

R. Y. Capangpangan, A. C. Alguno, *Microplastics in Focus*,
https://doi.org/10.1007/978-3-032-28978-0_10

10.1.1 From Concern to Credible Evidence: What Makes Community Monitoring Scientific

Community monitoring requires scientific discipline. Effective programs define clear target metrics, apply standardized collection and recording procedures, and integrate verification into each stage of the workflow [1]. For macro litter and plastic pellets, standardized shoreline surveys yield robust, comparable data across regions. A strong example comes from the Gulf of Mexico, where hundreds of trained volunteers applied a consistent sampling protocol to quantify pellet abundance and produce evidence that informed policy discussions and response strategies [4].

For microplastics research, the analytical bottleneck remains a critical constraint. Volunteers can collect environmental samples, yet laboratories must still isolate particles, identify them, and confirm polymer composition using techniques such as Fourier transform infrared spectroscopy or Raman spectroscopy. Reviews emphasize that limited laboratory capacity and unequal access to digital tools can deepen geographic and demographic disparities in participation [1]. Successful programs, therefore, align research questions with methods that communities can realistically sustain. Many grassroots initiatives begin with macro litter, pellets, and microplastics, then collaborate with accredited laboratories to confirm polymer type. Other teams focus on sample collection and preliminary sorting, while trained analytical hubs conduct polymer identification and quality assurance procedures.

Community-based monitoring operates not as isolated volunteer activities but as an integrated science-to-policy pathway connecting grassroots observation, scientific validation, institutional collaboration, and governance outcomes. Figure 10.1 illustrates this framework and shows how community participation can evolve into credible environmental evidence that informs policy decisions, strengthens environmental management, and generates societal benefits.

10.1.2 Monitoring Designs That Grassroots Groups Can Sustain

Grassroots monitoring often succeeds through designs that protect volunteer time. Three models recur across the literature.

1. *Shoreline and riverbank transects*: These designs use fixed routes and simple metrics, such as item counts, mass, or standardized pellet counts. They work well for identifying source zones, storm pulses, and cleanup priorities. Evidence shows that citizen science can reveal spatial and temporal drivers of litter patterns and land-use effects, thereby strengthening upstream prevention planning [1, 7].
2. *Watershed scale sampling*: A watershed design connects community sampling sites from headwaters to river mouths, enabling researchers to relate plastic

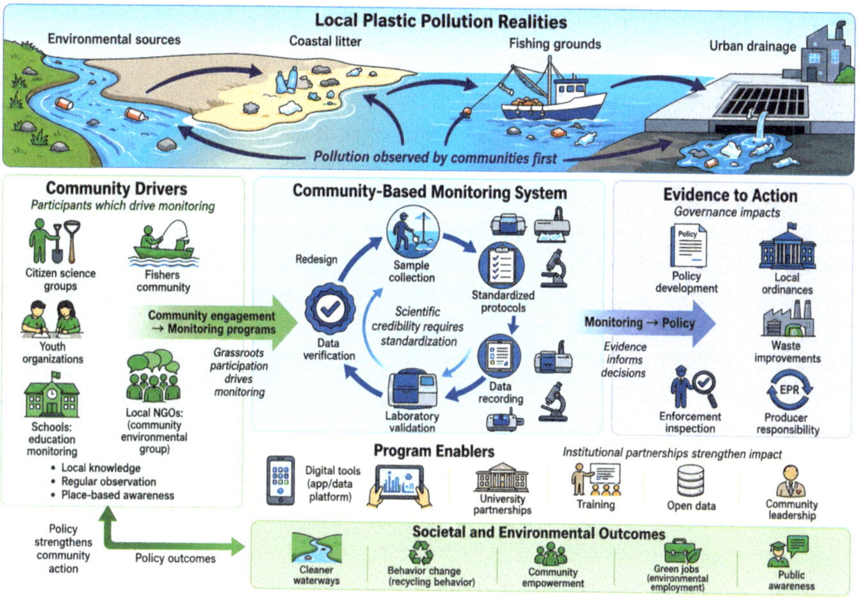

Fig. 10.1 Community-based microplastic monitoring as a science-to-policy pathway. (Conceptual framework developed by the authors and illustrated using digital design tools)

loads to land-use patterns and local activities. This approach clarifies how upstream practices influence downstream contamination. A watershed-scale citizen science study demonstrated that volunteers can effectively support microplastic concentration mapping when field protocols and laboratory processing are properly aligned [3].

3. *Nearshore or blue space sampling with trained use groups*: Some projects recruit "super users" such as surfers, sailors, or fishers who already spend time on the water. In the nearshore Mediterranean, citizen scientists conducting manta trawl-style sampling produced scientifically meaningful results on floating microplastic loads and seasonal variation [8]. This model can work in Southeast Asian coastal towns, where fishers and coastal youth groups often hold strong place-based knowledge and can sustain repeated sampling if programs provide simple gear and feedback.

10.1.3 Education as an Intervention: Youth-Led Monitoring That Changes Behavior

Grassroots monitoring does more than gather data. It reshapes how people interpret the litter around them. In an educational citizen science project in Italy, student participation measurably improved perceptions and knowledge about sources and transport of marine litter [6]. This matters for microplastics because prevention depends on daily behaviors: waste handling, consumption choices, and local advocacy. When schools adopt monitoring modules, students see plastic pathways from street drains to rivers to beaches, and they gain a practical reason to support local ordinances, better waste systems, and producer accountability.

10.1.4 Digital Tools, Open Data, and the Ethics of Participation

Digital data entry can strengthen participatory monitoring by reducing transcription errors, enabling photo verification, and improving feedback loops. Evidence from participatory environmental monitoring shows that digital data entry can support monitoring outcomes when projects design tools around user needs and local context [9]. Yet reviews warn about the digital divide, which can exclude communities with limited connectivity or devices [1]. Grassroots programs can respond with low-tech options: paper forms for later digitization, community data stewards, offline-capable apps, and shared devices for group surveys.

Ethics remains central to community monitoring. Projects can become extractive when researchers collect data without returning tangible benefits to participating communities. Strong programs communicate results promptly, involve local leaders as co-authors when appropriate, and align findings with community priorities, such as safer waste management, cleaner waterways, and equitable public services. Partnerships with frontline sanitation personnel further strengthen both scientific rigor and social equity, since these workers understand local waste flows and face direct exposure to plastic pollution risks [10].

10.1.5 How Grassroots Data Moves Policy

Policy relevance requires more than maps. Decision makers look for consistent methods, clear uncertainty statements, and a link to actions. The Nurdle Patrol case illustrates this pathway. Volunteers generated standardized pellet data across a large region, enabling the program to address the policy problem of pellet loss control and spill response. In Europe, citizen science datasets have also supported the evaluation of policy instruments aimed at reducing plastic pollution, demonstrating how community data can test the real-world effectiveness of these instruments [11].

At the global level, citizen science strengthens participation in plastic governance. Scholars argue that a global plastics treaty must "mind the people," recognizing that citizen science can expand public engagement and improve evidence flows, particularly in regions where formal monitoring systems remain limited [5]. This perspective aligns closely with realities in the Global South, where community groups often work with modest resources yet maintain strategic access to rivers, coastlines, and informal waste networks.

10.1.6 Practical Best Practices for Grassroots Monitoring Programs

The literature converges on a short list of practices that repeatedly improve outcomes.

1. *Start with a clear question and a small metric set*: Begin with a clearly defined research question and a focused set of metrics. Select indicators that volunteers can measure consistently and with confidence. Provide simple classification guides supported by clear photographs to improve accuracy and reduce observer error.
2. *Standardize protocols and training*: Provide short training, field practice, and refresher guides. Build inter-observer checks through paired surveys or periodic audits [1].
3. *Embed verification*: Incorporate photo uploads, automated range checks, and supervisor review to strengthen data validation. When projects report microplastic composition, link community sampling with laboratory confirmation of polymer identity to ensure analytical accuracy [8]. Share results promptly at the local level through community dashboards, barangay-level maps, and concise policy briefs that translate findings into practical actions, such as installing drain screens, improving market waste management, or implementing pellet containment measures.
4. *Build strong institutional linkages while safeguarding community leadership*: Strengthen institutional partnerships while preserving community leadership. Universities, local government units, and nongovernmental organizations can provide laboratory access, technical guidance, and data management systems, while community groups set priorities and co-design interventions that reflect local needs and realities [9, 10].

10.1.7 Why This Matters for Southeast Asia and Beyond

Southeast Asia faces significant plastic pollution in rivers and coastal waters. Rapid urbanization, expanding consumption, and uneven waste infrastructure intensify this challenge. At the same time, the region possesses strong civic networks, active youth movements, and responsive local government units that can mobilize monitoring initiatives at scale.

A growing body of peer-reviewed research on microplastics in mangroves, rivers, sediments, coastal waters, and urban air in the Philippines, Malaysia, Indonesia, Thailand, and Singapore provides a scientific foundation for community action. These studies provide locally relevant baselines to guide the development of context-specific protocols. Community monitoring in the region can therefore move beyond imported templates and adopt designs suited to tropical coastlines and dense urban watersheds. Examples include mangrove edge transects, river mouth hotspot mapping, port and fish landing site assessments, and market-to-drain waste pathway audits.

When communities integrate monitoring with upstream prevention strategies and active policy engagement, they transform observation into influence. Data generated at the barangay and watershed levels can inform ordinances, strengthen enforcement, and support producer accountability. This model demonstrates how Southeast Asia can position community science as a driver of regional leadership in plastic governance, with lessons that extend to other rapidly developing coastal regions worldwide.

10.2 Empowering Citizen Science in the Global South

Citizen science can move plastic monitoring from occasional cleanups to a credible evidence stream for decisions. This matters most in the Global South, where plastic leakage often concentrates near dense settlements, river corridors, and coastal livelihoods, yet formal monitoring remains thin. Well-designed citizen science can expand spatial coverage, improve temporal frequency, and surface local knowledge about sources and pathways that outsiders miss [1]. It can also strengthen policy traction when programs align with official indicators, transparent data standards, and clear quality controls [12].

10.2.1 Why Empowerment Is the Core Design Choice

Many projects invite participation but still keep power in distant institutions. Empowerment means communities shape the questions, control how results return to them, and share credit and benefits. Reviews warn that participation gaps can widen when programs depend on high-cost lab access, advanced digital tools, or English-only materials, because these barriers map onto existing inequities in education, infrastructure, and research funding [1]. A stronger model treats citizen science as shared governance. Communities provide site access, context, and continuity. Universities and agencies provide training, safety systems, and laboratory confirmation when needed.

10.2.2 Choose Questions That Match Feasible Methods

Microplastics create an "analytical bottleneck" because polymer confirmation requires techniques, such as FTIR or Raman, along with contamination control and careful QA/QC [1]. Programs, therefore, succeed when they match each question to a method that volunteers can deliver reliably. Many Global South initiatives start with macro litter, pellets, and mesoplastics because protocols are simpler, costs remain low, and results can still guide prevention actions, such as market waste rules, drain screens, or pellet loss controls [1, 13]. When projects target microplastics, they can split the workflow: communities collect and document samples, regional hubs perform density separation and subsampling, and partner laboratories confirm polymers using standardized spectral libraries and blank controls [1].

10.2.3 Build Quality into the Workflow, Then Make It Visible

Skeptics often dismiss citizen science on quality grounds. The solution is operational, not rhetorical. Programs can hardwire quality by using: (1) simple, photo-based classifications, (2) field duplicates, (3) automatic range checks in digital forms, (4) supervisor review, and (5) periodic intercomparison sessions across teams [1]. Reviews also emphasize that transparent QA/QC documentation increases scientific credibility and helps agencies integrate citizen science into formal reporting pathways. Skeptics often dismiss citizen science on quality grounds. The solution is operational, not rhetorical. Programs can hardwire quality by using: (1) simple, photo-based classifications, (2) field duplicates, (3) automatic range checks in digital forms, (4) supervisor review, and (5) periodic intercomparison sessions across teams [1]. Reviews also emphasize that transparent QA/QC documentation increases scientific credibility and helps agencies integrate citizen science into formal reporting pathways. This approach aligns with broader data stewardship guidance, such as the FAIR principles, which encourage datasets that others can find, access, and reuse with clear metadata and provenance [12].

10.2.4 Make Technology Serve Inclusion, Not Exclusion

Digital tools can widen participation when they reduce friction. Mobile platforms support geotagging, photo verification, and rapid dashboards. Examples include app-based litter logging and open databases that harmonize item lists across organizations, allowing pooled statistics across regions [14]. Open-source approaches can also lower costs and strengthen local ownership. For instance, citizen science-based tracking systems that rely on open hardware and software can support coastal litter observation without proprietary lock-in [15]. Still, programs in the Global South must plan for uneven connectivity. Offline-first data entry, multilingual interfaces, printable backup forms, and community data stewards can prevent dropout when phones fail or the signal disappears [1].

10.2.5 Link Citizen Science to SDG Reporting and Policy Cycles

Empowerment grows when data changes decisions. A clear pathway is SDG 14.1.1b, which tracks plastic debris density in coastal environments. Case discussions show that citizen science can support national reporting when agencies verify protocols and integrate datasets into official statistics, as reported for Ghana [16]. This linkage reframes community monitoring from "awareness activity" to "public evidence infrastructure." It also encourages stable funding, because SDG reporting creates recurring demand rather than one-time projects [17]. Programs can strengthen this pathway by adopting standardized shoreline survey designs, consistent effort measures, and open metadata that describe who sampled, when, where, and how.

10.2.6 Invest in People and Institutions, Not Only Kits

Many initiatives fail after the first grant cycle because they treat training as an event. Empowerment requires capacity that persists. Effective programs develop local trainers, certify team leads, and create small regional hubs that hold shared equipment, maintain reference materials, and coordinate sample shipment to partner laboratories [1]. This hub model also supports South-to-South learning, where neighboring countries exchange protocols suited to similar coastlines, monsoon cycles, and waste systems. Evidence from multi-country youth programs shows that structured participation can improve ocean literacy outcomes, but impacts vary with cultural context and implementation quality, so programs must evaluate learning and retention rather than assuming automatic benefits [18].

10.2.7 Address Ethics, Safety, and Benefit Sharing Directly

Plastic monitoring can expose volunteers to hazards such as contaminated waste, sharp objects, and unsafe shorelines. Programs should provide risk assessments, guidance on protective equipment, and clear rules for handling medical waste and hazardous items. Ethics also includes benefit sharing. Communities should receive rapid, plain-language results, local maps, and decision briefs, along with opportunities to co-author reports where feasible [1]. This matters in the Global South, where communities may experience "extractive data practices" that take observations without returning value or improving services. Partnerships with front-line waste workers and local government units can improve both validity and fairness because these stakeholders understand waste flows and face direct exposure risks in daily work [1].

10.2.8 A Practical Empowerment Blueprint for Global South Programs

1. *Co-define the problem*: Start with community priorities, such as market packaging, flood-driven waste pulses, river-to-coast leakage, or aquaculture gear loss.
2. *Select a small indicator set*: Use metrics that volunteers can measure consistently, then expand only after teams demonstrate reliability [1].
3. *Standardize protocols and training*: Use photos, short videos, and local language guides. Run periodic calibration sessions [13].
4. *Design hybrid validation*: Pair field data with hub-based QA/QC and laboratory confirmation when claims involve polymer identity [1].
5. *Publish usable outputs fast*: Provide barangay or district dashboards, hotspot maps, and short policy briefs that link results to feasible interventions.
6. *Institutionalize the partnership*: Anchor the program in universities, LGUs, and NGOs for data stewardship and lab access, while communities steer priorities and co-design actions.
7. *Keep data reusable*: Apply FAIR style metadata and transparent provenance so datasets remain credible beyond the project cycle [12].

The operational elements described above can be understood as an integrated empowerment pathway linking community participation, scientific validation, and governance uptake. Rather than functioning as isolated activities, these components form a connected system in which community knowledge, standardized monitoring workflows, quality assurance mechanisms, and institutional partnerships collectively transform citizen science into credible policy evidence. Figure 10.2 illustrates this science-to-policy pathway and highlights how empowered citizen science programs can bridge local monitoring efforts with national and global environmental governance.

The figure presents a conceptual framework illustrating how empowered citizen science in the Global South moves beyond participation to become a credible evidence system for environmental governance. The framework shows how community participation, standardized monitoring workflows, quality assurance systems, and institutional partnerships collectively transform local observations into scientifically robust data that support SDG reporting, local decision-making, and the institutionalization of long-term monitoring. As illustrated, empowered citizen science builds durable monitoring capacity where it is most needed, converts local

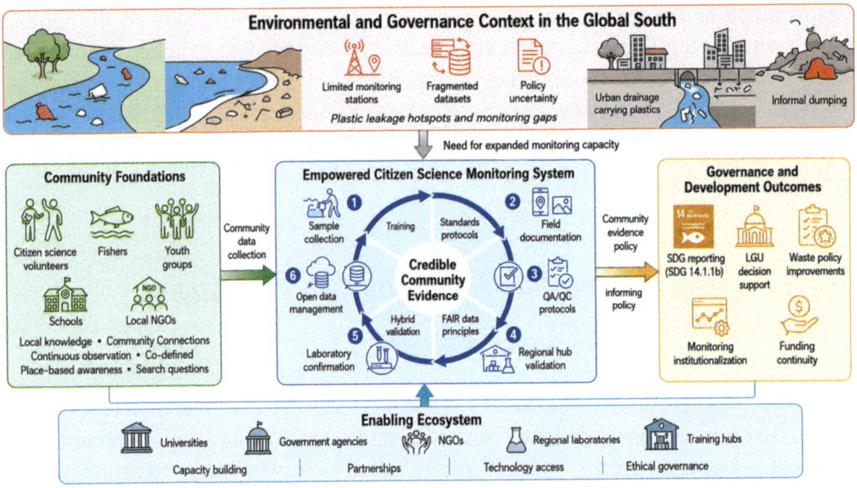

Fig. 10.2 Empowered citizen science as a bridge from community monitoring to policy impact in the Global South. (Conceptual framework developed by the authors and illustrated using digital design tools)

concerns into defensible evidence, and bridges community action to national and global plastic governance [16, 19].

10.3 Lifestyle Choices, Youth Advocacy, and Environmental Education

Lifestyle decisions shape plastic demand every day. When households buy, use, wash, discard, and replace products, they either reinforce disposable systems or signal a market shift toward reuse, refill, and safer materials. Social science evidence shows that people reduce plastic use when they understand risks, see practical alternatives, and face supportive cues, such as norms, convenience, and incentives [20]. Education and youth advocacy strengthen these cues by building skills, confidence, and collective identity around action and then translating awareness into community rules and purchasing pressure [21]. For microplastics, this pathway matters because major sources link directly to daily routines, especially textiles, packaging, and home waste practices [22, 23].

10.3.1 Start With Actions That Cut Release at the Source

Laundry stands out because synthetic textiles shed microfibers into wastewater. Community-scale evidence shows that washing machine filters can reduce microfiber emissions and can support policy discussions on source control [22]. Citizen science also demonstrates that households can install and use filters under real laundry conditions, while researchers track capture potential and social acceptability, which helps programs design interventions that people keep using [24]. These studies offer a practical education message: microplastics control can begin at home, but the solution must feel easy, affordable, and routine. Programs can teach a short set of laundry habits that reduce shedding, including full loads, lower agitation cycles, and careful lint handling, then pair these steps with filter pilots where feasible [23, 24].

10.3.2 Food and Shopping Routines Also Respond to Well-Designed Behavior Tools

Evidence from behavioral experiments shows that targeted interventions can reduce the selection of single-use produce bags. Effective designs include normative messages, attention cues that highlight the no bag option, and indirect incentives that connect choices to environmental outcomes. This matters for youth and education because it gives teachers and advocates concrete levers that work in real settings, such as school canteens, campus stores, wet markets, and groceries. A youth group can run a two-week produce bag challenge, track bag counts, and then present the results to store managers and local councils with a clear request, such as defaulting to no bag, visible prompts, or a small reward program [25].

10.3.3 Awareness Alone Does Not Guarantee Behavior Change, So Education Must Teach Decision Skills

Reviews of plastic-related perceptions and interventions highlight gaps between concern and action, often driven by convenience, uncertainty about what works, and weak infrastructure support [20]. Public surveys also show that many people support policies to reduce microplastics, yet they remain unfamiliar with the term and face barriers, such as limited information sharing and poor waste services [26]. Education can respond by focusing on three competencies: (1) source literacy, meaning where microplastics come from in daily life; (2) choice architecture, meaning how to make low plastic choices easier; and (3) civic pathways, meaning how to turn personal habits into shared rules and institutional procurement changes [20, 26].

10.3.4 Youth Advocacy Converts Learning into Policy Traction

A documented case of teen activism shows how a student group moved from classroom exposure to local laws banning single-use plastics through sustained organizing, public communication, and engagement with decision makers [21]. This model fits microplastics education because it treats students as producers of solutions rather than recipients of facts. It also fits Global South contexts where youth networks often operate through schools, faith groups, community councils, and local government units. The key lesson is strategic sequencing: define a local plastic problem that people recognize, run a measurable campaign, build alliances, then propose a policy change that institutions can enforce [21, 27]. Advocacy research also notes that civil society pressure in the Global South is increasingly elevating justice and equity concerns in plastics governance, including at the treaty level [27]. Youth movements can align with this justice framing by centering community health, sanitation worker safety, and the burden of unmanaged waste on informal settlements.

10.3.5 Design Environmental Education To Be Hands-On, Local, and Repeated

Evidence from a microplastics learning module with primary students shows that hands-on outreach activities can increase knowledge, and that both in-school and out-of-school settings can work when the module remains consistent and engaging [28]. Studies with university students also show that the mass media shape understanding of microplastics, but knowledge often remains partial and fragmented. This pattern argues for structured curricula that connect media claims to scientific reasoning and local context [29]. Formal education can therefore do three things well: (1) provide correct core concepts, including size classes, sources, and exposure pathways; (2) train students to evaluate evidence quality; and (3) link learning to action projects that produce local data or measurable waste reduction [28, 29].

10.3.6 Link Education to Health Framing Carefully and Ethically

Environmental health education can raise awareness of microplastic pollution among secondary students, supporting the inclusion of microplastics in school health and science programming [30]. In practice, educators should avoid fear messaging and focus on agency. Teach students to reduce unnecessary plastic contact where alternatives exist, while also explaining that upstream solutions, industry redesign, and governance drive the largest reductions. This balance protects students from anxiety and keeps the message aligned with evidence-based prevention [20, 30].

10.3.7 *Measure What Matters and Report It Back to Learners*

Strong programs track behavior outcomes, not only attitudes. Examples include counts of bags avoided, refill uptake, litter audits, microfiber filter adoption, and policy commitments from schools or shops. Where projects address microfibers, use simple indicators such as the number of participating households, filter maintenance rates, and estimated lint mass captured, then connect findings to a broader source reduction strategy [22, 24]. When programs report results through school dashboards, barangay briefs, and campus procurement updates, they reinforce norms and sustain motivation.

10.3.8 *Relevance for Southeast Asia and the Wider Global Context*

Many communities face high plastic leakage alongside strong youth networks and fast-moving social media ecosystems. Education and youth advocacy can scale quickly when they use locally relevant targets, such as sachet packaging, market drainage pathways, textile washing habits, and school procurement. Public perception studies show that willingness to act can be high even when technical understanding remains uneven, so programs can gain traction when they pair clear actions with visible community benefits [26]. When youth groups link daily habits to institutional change, they build a bridge from household behavior to policy and from local action to global debates on plastics governance [21, 27].

10.4 Localized Engagement Models from Southeast Asia and Beyond

Localized engagement models translate global science into community-owned solutions. They respond to cultural norms, governance structures, waste infrastructure realities, and ecological contexts. In Southeast Asia, where plastic leakage intersects with dense coastal populations and river-dependent economies, engagement must integrate science, policy, and livelihood considerations. The outline of this volume highlights the importance of regional relevance, especially for the Global South and Southeast Asia. This section advances that direction by examining how place-based initiatives reshape microplastic governance.

10.4.1 River Basin to Reef Approaches

Many Southeast Asian cities discharge plastics into river systems, where they fragment into microplastics. Effective engagement, therefore, begins upstream. Studies of river deltas and urban waterways show that microplastics travel from wastewater outlets into estuaries and coastal sediments, underscoring the need for watershed-scale monitoring [31]. Localized programs in Indonesia and the Philippines now align barangay-level cleanups with river mouth hotspot mapping. These efforts combine volunteer sampling with laboratory validation, building trust in local data.

Evidence from coastal islands in the Indian Ocean shows that even remote coral islands accumulate macro-, meso-, and microplastics due to regional circulation patterns [32]. Southeast Asian coral reef communities draw from this insight by linking reef monitoring with coastal waste audits. Fisherfolk associations record the types of plastic debris during reef surveys, then coordinate with municipal waste offices to address the most prevalent items, such as sachets and food packaging.

10.4.2 Textile and Urban Microfiber Awareness Networks

Microfiber release from textiles represents a major primary microplastic source. Experimental washing trials show that garments shed between 175 and 560 fibers per gram, depending on fabric characteristics [33]. This finding informs localized engagement models in Malaysia and Singapore, where university researchers collaborate with housing communities. Residents receive guidance on washing practices, the use of filters, and the selection of garments.

Such initiatives draw on global estimates of primary microplastic sources, which identify synthetic textiles as a significant contributor [34]. When local campaigns translate these global statistics into neighborhood-scale impacts, participation increases. Schools incorporate microfiber microscopy sessions, allowing students to visualize fibers collected from laundry effluent. This hands-on model bridges scientific abstraction and daily life.

10.4.3 Policy-Anchored Public Participation in Malaysia

Malaysia offers a structured example of multidimensional engagement. Recent analyses highlight technological innovation, policy instruments, and public participation as complementary pillars in addressing marine microplastic pollution [35]. Detection sensors improved monitoring accuracy, while plastic bag restrictions reduced usage by nearly 30% in participating states. Public monitoring programs increased voluntary cleanup participation and strengthened awareness.

Localized engagement in this context does not operate independently from regulation. Instead, it reinforces enforcement by providing community-based evidence. When residents track shoreline litter trends and share data with municipal councils, compliance pressure increases. This model aligns technological capacity with civic accountability.

10.4.4 Waste Infrastructure and Informal Sector Integration

Southeast Asian cities often rely on informal waste collectors who collect recyclable plastics before they are processed formally. Engagement models that exclude these actors overlook critical expertise. Research on plastic leakage emphasizes that waste management gaps drive environmental accumulation [36]. Local initiatives in Vietnam and Thailand now integrate informal recyclers into monitoring programs. Waste workers document the categories of plastic entering drainage systems during heavy rainfall events. Municipal engineers then adjust trap placement and waste flow management.

This inclusive model enhances data quality while addressing occupational exposure. It also reflects broader recognition that microplastics stem from systemic inefficiencies rather than isolated behaviors.

10.4.5 Community Science and Standardization

Localized engagement must maintain methodological rigor. Variation in sampling techniques and reporting units often complicates comparisons across regions [37]. Southeast Asian university networks, therefore, promote harmonized shoreline survey protocols and shared data repositories. Collaborative workshops train students and local officials in standardized counting, contamination prevention, and quality assurance.

Such networks mirror global calls for unified monitoring frameworks [38]. When local datasets follow comparable formats, they strengthen national reporting and international negotiations.

10.4.6 Coral Triangle and Small Island Contexts

The Coral Triangle region represents both a biodiversity hotspot and a plastic accumulation zone. Studies document microplastic ingestion across trophic levels, from zooplankton to large filter feeders [39]. Small island communities respond by embedding microplastic modules into marine protected area programs. Reef wardens collect sediment samples during routine patrols, and local universities analyze polymer types.

This model situates microplastic surveillance within existing conservation structures. It avoids duplicating administrative burdens and ensures continuity beyond short-term grants.

10.4.7 Beyond Southeast Asia

Localized engagement principles extend to Latin America, Africa, and small island states. Remote Arctic studies reveal that ocean currents transport microplastics far from original sources [40]. This finding underscores the interconnected nature of regional action. When Southeast Asian municipalities reduce leakage, downstream and distant ecosystems benefit.

Similarly, European shoreline accumulation research demonstrates how standardized surveys generate robust comparative datasets [41]. Southeast Asian programs adapt these methods while modifying classifications to reflect dominant packaging types in regional markets.

10.4.8 Strategic Characteristics of Effective Localized Models

Across contexts, effective localized engagement models share several characteristics:

1. Clear ecological focus linked to visible community assets, such as mangroves, fisheries, or tourism beaches.
2. Integration with existing governance units, including barangays, village councils, or municipal committees.
3. Scientific validation through university or laboratory partnerships.
4. Feedback loops that translate findings into waste management adjustments or policy refinement.
5. Youth participation to sustain generational continuity.

These features reflect the structure proposed in this volume, in which science informs policy and policy activates community response.

The localized engagement models described above demonstrate how diverse community-based initiatives across Southeast Asia translate scientific knowledge into practical governance solutions. Although these models operate in different ecological and institutional contexts, they share common structural features linking local monitoring, community participation, scientific validation, and policy integration. Figure 10.3 synthesizes these representative localized engagement pathways and illustrates how place-based initiatives collectively contribute to scalable regional microplastic governance.

The figure presents six complementary engagement pathways that show how community participation, scientific monitoring, and governance mechanisms can be integrated to strengthen microplastic management at local and regional scales. The River Basin to Reef Engagement Model emphasizes a source-to-sea monitoring continuum linking community sampling in rivers and wastewater systems with estuarine hotspot mapping and downstream monitoring of coral reefs and fisheries, demonstrating how watershed collaboration can connect upstream waste management with coastal ecosystem protection. The Textile Microfiber Awareness Model highlights how citizen science education can translate scientific knowledge on microfiber pollution from laundry into behavioral change through improved garment choices, washing filters, and household practices. Meanwhile, the Policy-Anchored Public Participation Model demonstrates how aligning monitoring technologies, policy instruments, and citizen engagement can strengthen regulatory effectiveness and environmental compliance.

The figure also highlights governance inclusion and scaling pathways. The Informal Sector Integration Model demonstrates how integrating informal waste collectors with municipal systems and community monitoring can improve both waste management and environmental monitoring outcomes. The Standardization and Community Science Networks Model shows how harmonized protocols, university training, and regional databases can transform local monitoring into policy-relevant scientific evidence. Finally, the Scalable Regional Network Model illustrates how local initiatives can expand into national policies, ASEAN regional collaboration, and global plastics governance. Together, these models demonstrate that localized engagement serves as a bridge between science and policy by linking community action, behavioral change, inclusive governance, and standardized monitoring into coordinated systems that support effective and scalable microplastic governance.

10.4.9 Toward Scalable Regional Networks

As illustrated in Fig. 10.3, localized engagement does not imply isolation; rather, it provides the foundation for scalable regional collaboration. Regional networks such as ASEAN research collaborations can harmonize protocols, share sensor technologies, and coordinate coastal monitoring campaigns. When communities document microplastic baselines and track trends over time, they build leverage in international plastics negotiations.

Southeast Asia possesses the demographic vitality, academic capacity, and civic momentum to pioneer such models. By grounding engagement in local realities while aligning with global science, the region can shift from pollution hotspot to innovation leader.

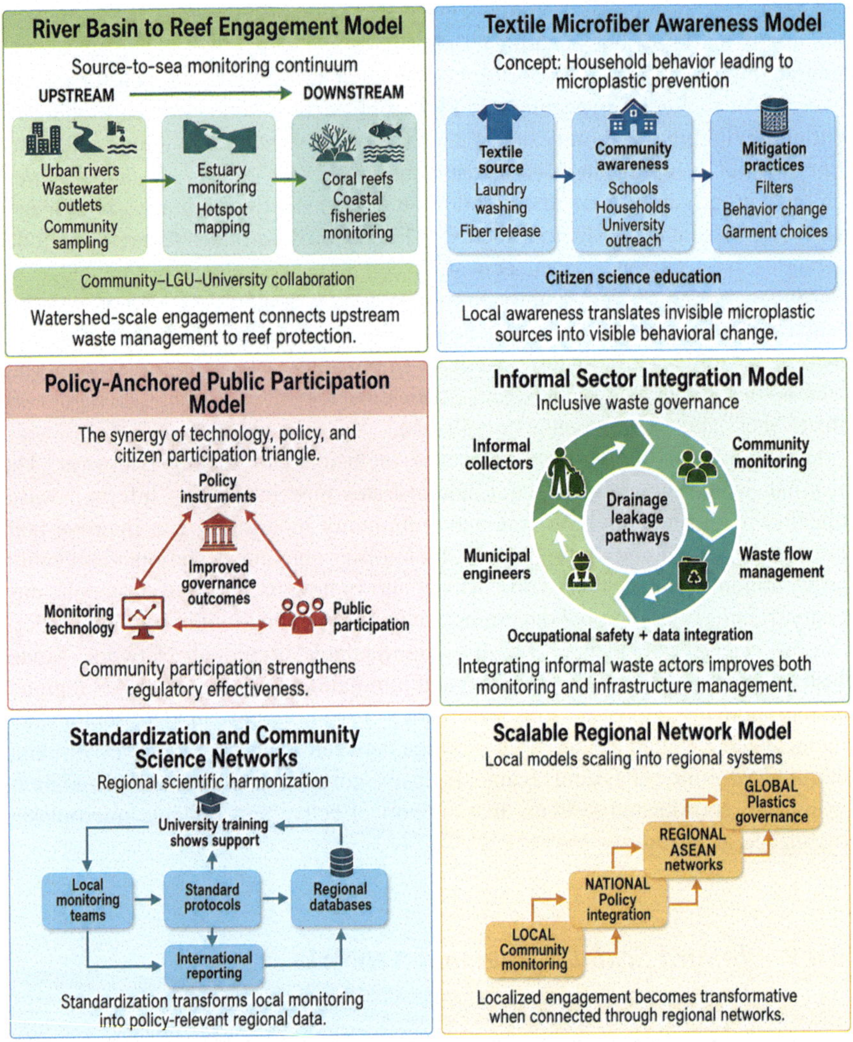

Fig. 10.3 Localized engagement models for microplastic governance in Southeast Asia and beyond. (Conceptual framework developed by the authors and illustrated using digital design tools)

10.5 Chapter Synthesis

This chapter demonstrates that effective responses to microplastic pollution extend beyond laboratories, regulatory agencies, and industrial systems. Communities, civil society organizations, educators, and individual citizens form a critical layer of environmental governance. Their participation expands monitoring coverage, strengthens environmental awareness, and builds the social momentum required to sustain long-term solutions. Evidence presented throughout this chapter shows that citizen science can generate credible datasets, strengthen transparency, and support policy development when supported by clear protocols, training, and institutional partnerships.

Grassroots monitoring initiatives illustrate how local observation can evolve into structured environmental intelligence. When volunteers follow standardized methods and collaborate with scientific institutions, community-generated data can complement formal monitoring programs and strengthen decision-making. These efforts also demonstrate that environmental stewardship begins with proximity. Communities often detect pollution trends earlier because they directly experience changing conditions in rivers, coastlines, and urban environments. Their observations, therefore, provide valuable early signals for prevention strategies and enforcement priorities.

Education and youth engagement further strengthen this foundation by transforming awareness into sustained behavioral change. Environmental education programs that connect scientific knowledge with daily practices encourage individuals to reduce plastic consumption, improve waste management habits, and advocate for responsible production systems. Youth movements demonstrate particular influence because they combine environmental literacy with civic participation. When educational initiatives integrate monitoring activities, advocacy training, and policy engagement, they create pathways from knowledge to measurable action.

The localized engagement models presented in this chapter further demonstrate that effective solutions must reflect social and environmental context. Programs across Southeast Asia demonstrate how monitoring can align with fisheries management, coastal conservation, urban waste planning, and school-based education. These approaches emphasize that environmental solutions become more durable when they reflect local governance structures, cultural practices, and livelihood realities. Community-based approaches, therefore, complement technological and policy interventions by strengthening ownership and accountability at the level where environmental change becomes visible.

This chapter also highlights an important structural insight. Long-term progress depends on partnerships that connect communities with universities, government agencies, and nongovernmental organizations. Such partnerships strengthen scientific credibility while ensuring that research outputs return tangible benefits to participating communities. When monitoring results translate into improved waste systems, local ordinances, and environmental education programs, community participation evolves from symbolic engagement into practical environmental governance.

This discussion reinforces a central lesson of the action agenda presented in this volume. Lasting solutions to microplastic pollution require alignment between scientific knowledge, technological innovation, policy direction, and societal participation. Communities provide continuity. Science provides evidence. Policy provides structure. Innovation provides tools. When these elements operate in coordination, environmental responses become both scalable and resilient.

References

1. Corbau, C., Lazarou, A., Bajt, O., Filipović Marijić, V., Simčič, T., Coltorti, M., Pignoni, E., & Simeoni, U. (2025). Citizen science for monitoring plastic pollution from source to sea: A systematic review of methodologies, best practices, and challenges. *Water, 17*(18), 2668. https://doi.org/10.3390/w17182668
2. Fraisl, D., Hager, G., Bedessem, B., Gold, M., Hsing, P.-Y., Danielsen, F., Hitchcock, C. B., Hulbert, J. M., Piera, J., Spiers, H., Thiel, M., & Haklay, M. (2022). Citizen science in environmental and ecological sciences. *Nature Reviews Methods Primers, 2*, 64. https://doi.org/10.1038/s43586-022-00144-4
3. Barrows, A. P. W., Christiansen, K. S., Bode, E. T., & Hoellein, T. J. (2018). A watershed-scale, citizen science approach to quantifying microplastic concentration in a mixed land-use river. *Water Research, 147*, 382–392. https://doi.org/10.1016/j.watres.2018.10.013
4. Tunnell, J. W., Dunning, K. H., Scheef, L. P., & Swanson, K. M. (2020). Measuring plastic pellet (nurdle) abundance on shorelines throughout the Gulf of Mexico using citizen scientists: Establishing a platform for policy-relevant research. *Marine Pollution Bulletin, 151*, 110794. https://doi.org/10.1016/j.marpolbul.2019.110794
5. Oturai, N. G., Syberg, K., Fraisl, D., Ramos, T. M., Schade, S., & Hansen, S. F. (2023). UN plastic treaty must mind the people: Citizen science can assist citizen involvement in plastic policymaking. *One Earth, 6*(6), 715–724. https://doi.org/10.1016/j.oneear.2023.05.024
6. Locritani, M., Merlino, S., & Abbate, M. (2019). Assessing the citizen science approach as tool to increase awareness on the marine litter problem. *Marine Pollution Bulletin, 140*, 320–329. https://doi.org/10.1016/j.marpolbul.2019.01.023
7. Vincent, A., Drag, N., Lyandres, O., Neville, S., & Hoellein, T. (2017). Citizen science datasets reveal drivers of spatial and temporal variation for anthropogenic litter on Great Lakes beaches. *Science of the Total Environment, 577*, 105–112. https://doi.org/10.1016/j.scitotenv.2016.10.113
8. de Haan, W. P., Uviedo, O., Ballesteros, M., Canales, Í., Curto, X., Guart, M., Higueras, S., Molina, A., Sanchez-Vidal, A., & The Surfing for Science Group. (2022). Floating microplastic loads in the nearshore revealed through citizen science. *Environmental Research Letters (Online), 17*(4), 045018. https://doi.org/10.1088/1748-9326/ac5df1
9. Brammer, J. R., Brunet, N. D., Burton, A. C., Currier, A., Danielsen, F., Dewan, K., Hermann, T. M., Jackson, M. V., Kennett, R., Larocque, G., Mulrennan, M., Pratihast, A. K., Saint-Arnaud, M., Scott, C., & Humphries, M. M. (2016). The role of digital data entry in participatory environmental monitoring. *Conservation Biology, 30*(6), 1277–1287. https://doi.org/10.1111/cobi.12727
10. Shruti, V. C., Pérez-Guevara, F., Roy, P. D., & Kutralam-Muniasamy, G. (2022). Strengthening citizen science partnerships with frontline sanitation personnel to study and tackle plastic pollution. *Environmental Science & Policy, 137*, 70–74. https://doi.org/10.1016/j.envsci.2022.08.011
11. Kiessling, T., Hinzmann, M., Mederake, L., Dittmann, S., Brennecke, D., Böhm-Beck, M., Knickmeier, K., & Thiel, M. (2023). What potential does the EU single-use plastics directive have for reducing plastic pollution at coastlines and riversides? An evaluation based on citizen science data. *Waste Management, 164*, 106–118. https://doi.org/10.1016/j.wasman.2023.03.042
12. Wilkinson, M. D., Dumontier, M., Aalbersberg, I. J., Appleton, G., Axton, M., Baak, A., Blomberg, N., Boiten, J.-W., Bonino da Silva Santos, L., Bourne, P. E., Bouwman, J., Brookes, A. J., Clark, T., Crosas, M., Dillo, I., Dumon, O., Edmunds, S., Evelo, C. T., …, Mons, B. (2016). The FAIR guiding principles for scientific data management and stewardship. *Scientific Data, 3*, 160018. https://doi.org/10.1038/sdata.2016.18
13. Kawabe, L. A., Ghilardi-Lopes, N. P., Turra, A., & Wyles, K. J. (2022). Citizen science in marine litter research: A review. *Marine Pollution Bulletin, 182*, 114011. https://doi.org/10.1016/j.marpolbul.2022.114011
14. Walker, T. R., McGuinty, E., & Hickman, D. (2021). Marine debris database development using international best practices: A case study in Vietnam. *Marine Pollution Bulletin, 173*(Part A), 112948. https://doi.org/10.1016/j.marpolbul.2021.112948

15. Merlino, S., Locritani, M., Guarnieri, A., Delrosso, D., Bianucci, M., & Paterni, M. (2023). Marine litter tracking system: A case study with open-source technology and a citizen science-based approach. *Sensors, 23*(2), 935. https://doi.org/10.3390/s23020935
16. Olen, S. M. (2022). Citizen science tackles plastics in Ghana. *Nature Sustainability, 5*, 814–815. https://doi.org/10.1038/s41893-022-00979-7
17. Gacutan, J., Oliver, J. L., Tait, H., Praphotjanaporn, T., & Milligan, B. M. (2022). Exploring how citizen science projects measuring beach plastic debris can support UN sustainable development goals. *Citizen Science: Theory and Practice, 7*(1), 34. https://doi.org/10.5334/cstp.459
18. Severin, M. I., Akpetou, L. K., Annasawmy, P., Asuquo, F. E., Beckman, F., Benomar, M., Jaya-Ram, A., Malouli, M., Mees, J., Monteiro, I., Ndwiga, J., Neves Silva, P., Nubi, O. A., Sim, Y. K., Sohou, Z., Tan Shau-Hwai, A., Woo, S. P., Zizah, S., Buysse, A., …, Catarino, A. I. (2023). Impact of the citizen science project COLLECT on ocean literacy and well-being within a north/west African and south-east Asian context. *Frontiers in Psychology, 14*, 1130596. https://doi.org/10.3389/fpsyg.2023.1130596
19. Ammendolia, J., & Walker, T. R. (2022). Citizen science: A way forward in tackling the plastic pollution crisis during and beyond the COVID-19 pandemic. *Science of the Total Environment, 805*, 149957. https://doi.org/10.1016/j.scitotenv.2021.149957
20. Heidbreder, L. M., Bablok, I., Drews, S., & Menzel, C. (2019). Tackling the plastic problem: A review on perceptions, behaviors, and interventions. *Science of the Total Environment, 668*, 1077–1093. https://doi.org/10.1016/j.scitotenv.2019.02.437
21. Burke, E., Ciano Collins, C., Bergeron, L., & Lavender Law, K. (2020). Teen activism leads to local laws banning single-use plastics: A two-year experiential learning journey. *Sustainable Earth, 3*, 15. https://doi.org/10.1186/s42055-020-00035-y
22. Erdle, L. M., Nouri Parto, D., Sweetnam, D., & Rochman, C. M. (2021). Washing machine filters reduce microfiber emissions: Evidence from a community-scale pilot in Parry Sound, Ontario. *Frontiers in Marine Science, 8*, 777865. https://doi.org/10.3389/fmars.2021.777865
23. Sheikhi, M., Lupato, S., Bianco, C., Sethi, R., & Tiraferri, A. (2024). Plastic microfibers from household textile laundering: A critical review of their release and impact reduction. *Critical Reviews in Environmental Science and Technology, 54*(20), 1501–1525. https://doi.org/10.1080/10643389.2024.2329513
24. Abourich, M., Bellamy, A., Ajji, A., & Claveau-Malette, D. (2024). Washing machine filters to mitigate microplastics release: Citizen science study to estimate microfibers capture potential and assess their social acceptability. *Environmental Challenges, 16*, 100984. https://doi.org/10.1016/j.envc.2024.100984
25. Luo, Y., & Zhao, J. (2023). Using behavioral interventions to reduce single-use produce bags. *Resources, Conservation and Recycling, 193*, 106942. https://doi.org/10.1016/j.resconrec.2023.106942
26. Praveena, S. M. (2024). Exploring public awareness, influencing factors and policy implications towards microplastic pollution: Perspectives from Malaysia. *Marine Policy, 161*, 106042. https://doi.org/10.1016/j.marpol.2024.106042
27. Dauvergne, P. (2023). Governing plastics: The power and importance of activism in the global south. *Environmental Science & Policy, 147*, 147–153. https://doi.org/10.1016/j.envsci.2023.06.011
28. Raab, P., & Bogner, F. X. (2021). Knowledge acquisition and environmental values in a microplastic learning module: Does the learning environment matter? *Studies in Educational Evaluation, 71*, 101091. https://doi.org/10.1016/j.stueduc.2021.101091
29. Janoušková, S., Teplý, P., Fatka, D., Teplá, M., Cajthaml, T., & Hák, T. (2020). Microplastics—How and what do university students know about the emerging environmental sustainability issue? *Sustainability, 12*(21), 9220. https://doi.org/10.3390/su12219220
30. Bayoğlu, M. A., & Uncu, F. (2024). The effect of environmental health education on microplastic pollution awareness. *Public Health Nursing, 41*(4), 760–767. https://doi.org/10.1111/phn.13322
31. Leslie, H. A., Brandsma, S. H., van Velzen, M. J. M., & Vethaak, A. D. (2017). Microplastics en route: Field measurements in the Dutch river delta and Amsterdam canals, wastewater

treatment plants, North Sea sediments and biota. *Environment International, 101*, 133–142. https://doi.org/10.1016/j.envint.2017.01.018

32. Imhof, H. K., Sigl, R., Brauer, E., Feyl, S., Giesemann, P., Klink, S., Leipold, K., Löder, M. G. J., Löschel, L. A., Missun, J., Muszynski, S., Ramsperger, A. F. R. M., Schrank, I., Speck, S., Steibl, S., Trotter, B., Winter, I., & Laforsch, C. (2017). Spatial and temporal variation of macro-, meso- and microplastic abundance on a remote coral Island of the Maldives, Indian Ocean. *Marine Pollution Bulletin, 116*(1–2), 340–347. https://doi.org/10.1016/j.marpolbul.2017.01.010
33. Belzagui, F., Crespi, M., Alvarez, A., Gutierrez-Bouzán, C., & Vilaseca, M. (2019). Microplastics' emissions: Microfibers' detachment from textile garments. *Environmental Pollution, 248*, 1028–1035. https://doi.org/10.1016/j.envpol.2019.02.059
34. Boucher, J., & Friot, D. (2017). *Primary microplastics in the oceans: A global evaluation of sources*. International Union for Conservation of Nature (IUCN). https://doi.org/10.2305/IUCN.CH.2017.01.en
35. Goh, K. C., Kurniawan, T. A., Sarpin, N., Masrom, M. A. N., Othman, M. H. D., Anouzla, A., Aziz, F., Ali, I., Casila, J. C., Khan, M. I., Zhang, D., Mahmoud, M., & Onn, C. W. (2025). Combating microplastic pollution in Malaysia's marine ecosystems using technological solutions, policy instruments, and public participation: A review. *Journal of Hazardous Materials Advances, 17*, 100542. https://doi.org/10.1016/j.hazadv.2024.100542
36. Jambeck, J. R., Geyer, R., Wilcox, C., Siegler, T. R., Perryman, M., Andrady, A., Narayan, R., & Lavender Law, K. (2015). Plastic waste inputs from land into the ocean. *Science, 347*(6223), 768–771. https://doi.org/10.1126/science.1260352
37. Horton, A. A., Walton, A., Spurgeon, D. J., Lahive, E., & Svendsen, C. (2017). Microplastics in freshwater and terrestrial environments: Evaluating the current understanding to identify the knowledge gaps and future research priorities. *Science of the Total Environment, 586*, 127–141. https://doi.org/10.1016/j.scitotenv.2017.01.190
38. GESAMP. (2015). *Sources, fate and effects of microplastics in the marine environment: A global assessment*. IMO/FAO/UNESCO-IOC/UNIDO/WMO/IAEA/UN/UNEP/UNDP Joint Group of Experts on the Scientific Aspects of Marine Environmental Protection.
39. Fossi, M. C., Coppola, D., Baini, M., Giannetti, M., Guerranti, C., Marsili, L., Panti, C., de Sabata, E., & Clò, S. (2014). Large filter feeding marine organisms as indicators of microplastic in the pelagic environment: The case studies of the Mediterranean basking shark (*Cetorhinus maximus*) and fin whale (*Balaenoptera physalus*). *Marine Environmental Research, 100*, 17–24. https://doi.org/10.1016/j.marenvres.2014.02.002
40. Peeken, I., Primpke, S., Beyer, B., Gütermann, J., Katlein, C., Krumpen, T., Bergmann, M., Hehemann, L., & Gerdts, G. (2018). Arctic Sea ice is an important temporal sink and means of transport for microplastic. *Nature Communications, 9*, 1505. https://doi.org/10.1038/s41467-018-03825-5
41. Browne, M. A., Crump, P., Niven, S. J., Teuten, E., Tonkin, A., Galloway, T., & Thompson, R. (2011). Accumulation of microplastic on shorelines worldwide: Sources and sinks. *Environmental Science & Technology, 45*(21), 9175–9179. https://doi.org/10.1021/es201811s

Chapter 11
Conclusion and Future Direction

This book has examined microplastics from their molecular origins to their global consequences and the collective responses now emerging to address them. It has shown how plastic production, fragmentation, and environmental transport generate persistent particles that circulate through oceans, rivers, soils, the atmosphere, and food systems. It has examined ecological disruption, toxicological concerns, economic implications, and governance gaps that define the current environmental reality. It has also presented practical pathways for change through material redesign, circular economy transitions, improved waste systems, sector-specific mitigation in agriculture and aquaculture, technological innovation, and community-driven monitoring. A central conclusion emerges clearly. Microplastic pollution represents a systems challenge that connects science, industry, governance, and human behavior.

11.1 Turning the Tide: Integration of Science, Policy, and Action

The chapters of this book follow the scientific journey from formation to consequence to solution. Early discussions clarified how plastic materials fragment through mechanical stress, ultraviolet radiation, chemical weathering, and biological activity, producing persistent micro- and nanoscale particles capable of long-distance transport across environmental compartments. These processes link production systems, consumption patterns, and waste management performance with environmental exposure.

Subsequent discussions examined ecological and human exposure dimensions. Microplastics accumulate in sediments, organisms, and food webs. They influence soil structure, affect the physiology of aquatic organisms, and act as carriers of chemical additives and environmental contaminants. Their detection in drinking water, seafood, and human tissues highlights a complex exposure network that

R. Y. Capangpangan, A. C. Alguno, *Microplastics in Focus*,
https://doi.org/10.1007/978-3-032-28978-0_11

intersects environmental health, food security, and social vulnerability. Economic implications also emerge through impacts on fisheries productivity, tourism sustainability, agricultural soil quality, and the financial burden of waste management systems.

Later sections shifted the focus toward solutions. Technological innovation now provides intervention points across the plastic life cycle. These include safer material design, biodegradable polymers supported by environmental validation, circular economy frameworks, advanced wastewater treatment technologies, nanotechnology-based capture systems, and targeted mitigation approaches in high-leakage sectors, such as agriculture and aquaculture. Community science initiatives further expand monitoring capacity and strengthen environmental accountability by connecting scientific knowledge with local observation and governance.

A consistent principle runs through these discussions. Effective control of microplastic pollution depends on integration. Scientific evidence must guide standards and monitoring frameworks. Policy must create enforceable accountability. Industry must redesign production systems. Communities must participate in prevention and monitoring. Fragmented responses cannot address a pollutant that crosses environmental boundaries and governance jurisdictions.

11.2 Strategic Positioning: Microplastics as a Defining Environmental Governance Challenge

Microplastic pollution now occupies a critical position within global sustainability discussions because it reflects broader structural challenges in modern production systems. It reveals limitations of linear economic models, weaknesses in waste infrastructure, and gaps in environmental accountability. At the same time, it presents an opportunity to redesign materials, strengthen environmental governance, and accelerate transitions toward sustainable production systems.

This issue also represents an emerging frontier in environmental science. Future research must strengthen the understanding of nanoplastic behavior, long-term toxicological pathways, trophic transfer mechanisms, atmospheric transport dynamics, and interactions with climate-related stressors. Advances in analytical chemistry, remote sensing, artificial intelligence-assisted monitoring, and predictive modeling will expand detection capacity and risk assessment precision. Harmonized monitoring standards and open data systems will be essential for ensuring comparability across regions.

The policy landscape will also evolve. Extended producer responsibility frameworks, plastic treaty negotiations, sustainable materials standards, and lifecycle accountability systems will likely shape the next generation of environmental regulation. These developments require strong scientific foundations to ensure that interventions remain evidence-based and measurable.

11.3 Global Partnerships and the Leadership Role of the Global South

Microplastic pollution operates as a transboundary challenge. Rivers transport plastic across countries. Ocean currents redistribute particles across continents. Atmospheric transport expands their reach beyond traditional pollution boundaries. This reality requires coordinated international responses supported by shared monitoring protocols and aligned regulatory frameworks.

The Global South holds a particularly important role in shaping future responses. Rapid urban growth, expanding coastal populations, and infrastructure gaps increase exposure risks in many regions of Southeast Asia, Africa, and Latin America. At the same time, these regions demonstrate strong innovation capacity through community-based monitoring, informal recycling networks, and adaptive environmental governance approaches.

Future progress depends on strengthening scientific capacity in these regions through laboratory development, training programs, access to technology, and equitable research partnerships. Collaboration must move beyond sample collection toward co-creation of knowledge, shared authorship, and reciprocal capacity development. Such partnerships strengthen both scientific quality and policy relevance.

Regional cooperation across Southeast Asia in particular offers strong potential for advancing harmonized monitoring standards, coordinated marine protection strategies, and shared technological innovation. Investments in waste infrastructure, inclusive recycling economies, and circular material systems can significantly reduce leakage while supporting sustainable livelihoods.

11.4 Future Directions: Science, Innovation, and Societal Transformation

The next decade will determine whether microplastic pollution stabilizes or continues to expand. Progress will depend on several priority directions.

Research must advance toward understanding long-term exposure risks, especially chronic low-dose exposure and nanoplastic interactions with biological systems. Standardized analytical protocols must improve data comparability. Monitoring must expand into underrepresented regions and environmental compartments.

Innovation must continue to redesign materials for durability, recyclability, and environmental compatibility. Circular economy models must move from pilot projects toward mainstream industrial practice. Waste systems must integrate prevention, recovery, and redesign rather than relying solely on downstream cleanup.

Governance must evolve toward lifecycle accountability. Policies must address production design, consumption patterns, and post-use recovery simultaneously.

International cooperation must strengthen implementation capacity rather than focusing solely on commitments.

Education and behavioral science will also play decisive roles. Environmental literacy programs must connect individual consumption decisions with systemic environmental outcomes. Youth engagement will remain particularly important because generational transitions often drive environmental transformation.

11.5 Inspiring Hope: Toward Resilient, Plastic-Conscious Futures

Microplastics represent a challenge created by human innovation. They also present an opportunity to demonstrate how science, governance, and society can respond to complex environmental risks. Advances in materials science, environmental monitoring, waste engineering, and sustainability governance show that alternatives exist. Community engagement and youth advocacy demonstrate that societal values can shift toward environmental responsibility.

Progress will depend on measurable actions. Each improved product design, each upgraded waste facility, each validated monitoring dataset, and each science-based policy contributes to cumulative environmental improvement. Transparency, interdisciplinary cooperation, and accountability across supply chains will determine the pace of this transition.

The future trajectory of microplastic pollution will depend on whether societies choose prevention over reaction, redesign over disposal, and cooperation over fragmentation. Scientific knowledge now provides sufficient clarity to guide these choices.

11.6 Final Reflection: Science, Responsibility, and the Future We Choose

Microplastics have revealed a difficult truth about modern progress. Materials designed for convenience have outlived their usefulness and entered the deepest layers of natural systems. They circulate through ocean gyres, settle into river sediments, accumulate in agricultural soils, and travel through the air that sustains life. Their presence in food systems and human tissues signals a new phase of environmental awareness, one in which pollution is measured not only by what can be seen but also by what persists beyond immediate perception.

Yet this challenge also demonstrates the strength of scientific progress. Within a few decades, research has moved from discovery to detection, from detection to risk assessment, and from risk assessment to solution design. New materials are being developed. Monitoring tools continue to improve. Policies are evolving. Communities are becoming active participants in environmental protection.

The direction of the next decades will depend on whether these advances remain fragmented or become coordinated. Environmental recovery will not depend on a single discovery. It will depend on whether knowledge systems, governance structures, industries, and societies move with a shared purpose. Microplastics, therefore, represent more than pollution. They represent a test of whether science can guide humanity toward more responsible forms of development.

11.7 Why This Book Matters

This book was written to serve as a bridge between knowledge and response. It brings together scientific evidence, technological pathways, governance strategies, and societal approaches within a single integrated framework. Its purpose extends beyond describing a problem. It seeks to clarify how scientific understanding can inform practical solutions across disciplines and sectors. By presenting the full continuum from origin to impact to intervention, it provides a foundation for researchers developing new studies, policymakers designing regulatory frameworks, educators building environmental literacy, and practitioners implementing mitigation strategies. At a time when information often remains fragmented across disciplines, this volume offers a structured synthesis intended to support coordinated action. Its value, therefore, lies in integration. It connects evidence with application and analysis with implementation.

11.8 A Legacy Perspective

Books on environmental challenges often serve as markers of scientific maturity. They capture what a generation has learned and define the questions that the next generation must answer. This volume aims to contribute to that tradition. It documents the current scientific understanding of microplastics while emphasizing the need for continued vigilance, innovation, and cooperation.

The long-term significance of this work will not be measured only by citations or academic discussion. Its true value will be measured by how it contributes to better monitoring systems, improved material design, stronger environmental policies, and more informed communities. If this book helps strengthen interdisciplinary research, supports evidence-based decisions, and encourages responsible environmental stewardship, then it will have fulfilled its purpose.

Scientific progress always builds across generations. The responsibility of the present is to ensure that the knowledge produced today reduces the environmental burdens inherited by the future.

The story of microplastics is still being written, and its final outcome will depend on whether humanity chooses to let plastic define the future of the environment or whether science, responsibility, and collective action will define the future of plastic.

Index

R. Y. Capangpangan, A. C. Alguno, *Microplastics in Focus*,
https://doi.org/10.1007/978-3-032-28978-0

The manufacturer's authorised representative in the EU is Springer Nature Customer Service Centre GmbH, Europaplatz 3, 69115 Heidelberg, Germany. If you have any concerns regarding our products, please contact ProductSafety@springernature.com

Printed and bound by CPI Group (UK) Ltd, Croydon, CR0 4YY
09/10/2026
02230868-0001